Enterprise Network Performance Optimization

Martin Nemzow

Warranty for included CD-ROM is
found on the last page of this book.
The material on the CD-ROM may
not be distributed without the book.

McGraw-Hill, Inc.

New York San Francisco Washington, D.C. Auckland Bogotá
Caracas Lisbon London Madrid Mexico City Milan
Montreal New Delhi San Juan Singapore
Sydney Tokyo Toronto

Other Books by Martin Nemzow

Keeping the Link

The Ethernet Management Guide

The Token-Ring Management Guide

FDDI Networking

LAN Performance Optimization

Computer Performance Optimization

Titled in the McGraw-Hill Series on Computer Communications

PETERSON • *TCP/IP Networking: A Guide to the IBM Environment*

DAVIS, McGUFFIN • *Wireless Local Area Networks: Technology, Issues, and Strategies*

HUTCHINSON • *ISDN Computer Applications Development*

SIGNORE, CREAMER, STEGMAN • *The ODBC Solution: Open Database Connectivity in Distributed Environments*

RICHARDSON • *Writing VX-REXX Programs*

SCHATT • *Linking LANs—2nd Edition*

BLACK • *TCP/IP and Related Protocols—2nd Edition*

BLACK • *Network Management Standards: OSI, SNMP, and CMOL Protocols—Second Edition*

FATAH • *Electronic Mail Systems: A Network Manager's Guide*

*For more information about other McGraw-Hill materials,
call 1-800-2-MCGRAW in the United states. In other
countries, call your nearest McGraw-Hill office.*

Special thanks to all those people and organizations who provided help for this book, particularly to Carol Weingrod, and my children, Sophie, Esther, and Gabriel, who wanted to see the work that Daddy was doing in the middle of the night.

Marty Nemzow
May 1994

Library of Congress Cataloging-in-Publication Data

Nemzow, Martin A. W.
    Enterprise network performance optimization / by Martin Nemzow.
        p. cm.
    Includes bibliographical references and index.
    ISBN 0-07-911889-5
    1. Local area networks (Computer networks)—Management.
    2. Business—Data processing.   I. Title.
    TK5105.7.N447 1994   94-3427
                                                                    CIP

Enterprise Network Performance Optimization
ISBN 0-07-911889-5
The material contained here is a complete published book and the hypertext
help file included with the book. This represents the entirety of the perfor-
mance analysis and optimizing techniques provided in the complete book.
Most networks and computing platforms are represented from Macintosh, to
PCs, to mainframes. All references and sources are included.

*The sponsoring editor of this book was Brad Schepp and the book editor
was Aaron Bittner. The pages were produced by the author.*

*In order to receive additional information on these or any other
McGraw-Hill titles, in the United States please call 1-800-822-8158.*   *9118895*
In other countries, contact your local McGraw-Hill representative.   MH94

Contents

Acknowledgments   **xi**

Introduction   **xiii**

In a few words.... xiii
Acknowledgments to readers and contributors xv

Chapter 1   Overview   **1**

Introduction 1
Purpose of book.. 5
Intended audience... 6
Book content... 7
Purpose for optimization 7
About the included reference CD-ROM 8
Structure of book 8

Chapter 2   Performance Paradigms   **11**

Introduction 11
Transportation system traffic... 12
           Traffic dynamics 13
           Enterprise network infrastructure 23
Network enterprise aspects.. 24
           Channel capacity 25
           Bandwidth 25
           Signal speed 26
           Route 26
           Protocol overhead 26
           Delay (latency)... 27
           Accidental latency 28
           Peaks and bursts 29
           Frame size 33
           Volume... 33
Data communication highway.. 34
           Mechanical infrastructures... 35

Leaks and losses..37
Pressure and volumes...38
LANs 40
Manageable operations 44
Firewalls...45
MANs 45
Load and latency limitations...45
WANs 46
Bandwidth limitations..46
Costs 47
Shared media vs. dedicated bandwidth 47
Workload contribution 48
Workload characterization 48
Scalability...49
Transaction processing 50
Backplanes vs. hubs 52
Integration 52
Conclusion...53

Chapter 3   Bottlenecks   55

Introduction 55
Definition of bottleneck...55
Establish performance service levels 59
Top hits...59
Bandwidth and latency 61
Bandwidth and bandwidth utilization 61
Mobile and wireless bandwidth 65
Memory bandwidth..65
Latency...66
Mobile and wireless latency..71
Network load 71
Shared media 74
Bandwidth myth...75
Interconnectivity devices...75
NICs 77
MAU or wiring concentrator 77
Repeater 78
Hub...78
Bridge...79
Router...80
Multiport router...81
Gateway 83
Switch 84
Backbone interconnection 86
Protocol differences...88
Multiprotocol stacks 90
Network protocols..90

Slow FTP 91
A disconnected device 91
Ascertaining true causes for bottlenecks 92
Rightsizing and control of distributed systems 93
    Remote and modem communications 94
    Remote agents for clients and connectivity 96
    Limit access to resources 96
Management overhead 97
    Voice over the network 97
    Imaging and video 98
    Fax imaging 99
Conclusion 99

Chapter 4   Tools   101

Introduction 101
Performance tools are few 102
    Design 103
Blueprinting 103
    Geographical views 103
    Wiring route map 104
    Logical structure 105
    Equipment diagrams 106
    Process flows 107
Inventory tracking 110
    Spare parts 111
Disaster recovery planning 112
    Computer system and network performance measurement tools 115
Software metering 123
Activity tracking 124
Disk waste 125
    Rummage 126
Profilers 127
Update management 130
Cable certification 131
Protocol analysis 132
    How to apply protocol analysis 138
    Management statistics 139
Types of benchmarks 140
    Applying benchmarks 143
Workload characterization 144
Tuning tools 145
Modeling the enterprise network 145
    Extrapolation modeling 146
    Simulation modeling 147
    Statistical modeling 150
    Emulation modeling 153
    Capacity planning 155

Bridging vs. routing 155
Switching vs. routing 160
Router buffer size.. 161
Bandwidth and latency.. 162
Reliability and survivability.. 164
Simulation of packet size 167
LANModel transaction modeling 171
Enterprise network cross-product traffic 171
Conclusion.. 173

Chapter 5   Optimizing Networks   175

Introduction 175
Infrastructure fortification 176
Optimization methods 177
Optimizing with design 179
Generalized optimization techniques 195
Designing the enterprise network.. 195
     Purpose.. 196
     Wiring infrastructure.. 196
     Redundancy.. 197
     Security 198
     Integration.. 198
     Scalability 200
Financial optimization... 201
     Consolidation of duplicate resources 201
     Consolidation of transmission services..................................... 202
     Better productivity from people with better communications 202
     Decreased use of fax and meetings, and more reliance on networking 202
     Fax service optimization 203
     Optimize transmission costs 203
     Overlooked benefits of performance optimization 204
     Recycled equipment... 204
Technical solutions.. 205
     Reduce the workload.. 205
     Shift the paradigm.. 206
     Microsegmentation 207
     Decrease the transmission latency 208
     Collapse the backbone 209
     Increase the transmission bandwidth 210
     Bandwidth on demand 214
     NICs 214
     Repeater 215
     Gateway 215
     Bridge... 216
     Router... 217
     Routing algorithms... 218
     Microsegmentation devices 220

Hub.. 221
Buffered network I/O.. 222
Predictive pipelining and NIC enhancements.. 223
Virtual switching.. 223
Port switching 229
Full-duplex transmission... 229
Power bandwidth 230
Optimization details 231
Energy usage.. 231
110 Alert 232
Green Keeper 234
Optimizing desktops.. 234
Optimizing WANs... 234
Connecting LANs 235
Optimizing with multiprotocol stacks.. 235
Optimizing network operating systems 236
TCP/IP drivers 236
Multiple stacks 237
Large internet packets 238
Handicapped wait.. 238
Cache configuration 238
Tracking the disconnected device 241
Optimizing application code 241
Networks with synchronous protocols 245
Optimizing transaction processing 246
Multithreading and multitasking code optimization 247
Client/server transaction processing... 248
Tune application code 249
SQL... 249
Visual Basic 253
PowerBuilder 255
DB2 256
SNA gateway performance optimization.. 257
Replication services 259
Enterprise naming 261
Imaging optimization 261
Printer configurations.. 262
Servers.. 266
Clients... 268
NetWare DOS Requester 271
X Windows.. 272
Anomalies 273
Optimizing WAN connections... 273
Financial performance considerations for WAN links 274
Optimizing hub and spoke networks... 274
Instant MHS services 275
Improve local and remote performance 275
Conclusion.. 277

Chapter 6   The CD-ROM   **279**

Introduction 279
     Handling 279
CD-ROM contents 280
Installation 281
On-line hypertext 281
     Windows help format 281
     Multimedia Viewer format 283
     Computer Performance Optimization demonstration 287
Demonstrations 287
     GrafBase demonstration 288
     NET F/X 289
     Synoptics SNA workstation controller 291
     SimSoft 291
     CACI COMNET III 293
     ServerTrak and TrendTrak 294
     LANalyzer for Windows 295
     Prophesy 297
     SysDraw 297
     LAN•CAD 298
     netViz 299
     LANBuild 302
     Win, What, Where 303
     Rummage 305
     NetSpecs 306
     ScreenCam movies 308
     Morph design movies 310

Sources   **313**

Bibliography and references 313

Index   **321**

Acknowledgments

Special thanks particularly to the people who provided help for this book; Many are listed in the Bibliography. Brad Schepp was the acquisitions editor for this project at Windcrest and McGraw-Hill. Aaron Bittner completed my sentences and clarified some of my more obscure concepts. Special thanks to Wextech Systems, Inc., who provided Doc-To-Help, and to Paul Neshamkin and Virginia Wade for showing me the techniques for making it work well. Special thanks to LaserMaster Corporation and their WinJet adapter and drivers. Kudos to Bob Pratt, LAN Product Manager at Novell, who supplied LANalyzer for Windows, and adapted a special Token-Ring and Ethernet demonstration and slide show for this book. Microsoft supplied the Viewer Compiler, which created the hypertext on the CD-ROM so as to include the sound bites, 256-color graphics, and keyword search engine. Lotus contributed an early release of their ScreenCam video so that I could create some modeling and design demonstrations and overdub them with sound.

Marty Nemzow
May 1994

Introduction

In a few words...

Although *Enterprise Network Performance Optimization* was already envisioned while *LAN Performance Optimization* was still in production, several other assignments, commercial products, and McGraw-Hill books delayed this one. Also, many people needed to see that performance was a real issue for expanding LANs, campus, wide area, and global networks—and that this topic needed its own book. Unfortunately, during this time the necessity for configuring and tuning large networks has become that much more acute. However, at least I can tell you about the new methods, tools, and opportunities suddenly available for optimizing an enterprise network. The new technologies of network virtualization, dedicated bandwidth, switching, and an improved and supportive atmosphere for component intercompatibility have supplemented the traditional mechanical microsegmentation of shared media networks. Once you get beyond the vendor hype to understand the design issues and configuration limitations, there are more available methods and economically-feasible ways to improve performance. This is very important to those of you with large bases of Intel 286-based PCs, 3270 terminals, and legacy equipment with small chance of being upgraded this century.

More technical choices do not necessarily mean that optimization is easier, cheaper, quicker, or voted more likely to succeed. These choices may not even represent viable options for your situation or correct solutions to enterprise performance bottlenecks. Enterprise networking represents data communications, information processing, flows of complex, sequenced, and integrated tasks potentially separated by many different conduits for networking, all sorts of different boundary layers, and many different methods for information storage and processing. Note that the primary enterprise network transmission protocols are LAN-based protocols, such as Ethernet, Token-Ring, or FDDI. Links between remote sites are primarily established with WAN-based protocols over private, dedicated, or switched links. You probably know these telecommunication protocols and carrier methods as SVC, PVC, DS0, frame relay, and ISDN. There is no true "enterprise protocol" other than IPX/SPX, TCP/IP, or ISO, although ATM carries an as yet unfulfilled pledge that it can provide immense scalable bandwidth to integrate voice, video, and data transport for LANs, WANs, and enterprises. As such, the enterprise network designer, builder, and manager must work within the constraints of transmission protocols and carrier methods.

Most of the performance problems that plague an enterprise network are likely to be caused by network management oversights, improper network management activities, bad configuration of key network components, poor hardware and software integration, junk traffic, badly designed distributed (or client/server) applications, and poor network design. Too many applications are ill-conceived and irreconcilable with WAN bandwidth and routing limitations. It can be *that* simple. However, performance problems can be temporarily disguised by faster processors, different software, more bandwidth, integration or implementation obstacles, and Star Wars solutions. Then the performance problem is mired in political muck and finger-pointing. Believe *that*.

Enterprise network performance optimization is not always about replacing sluggish Ethernets with faster protocols, such as 100Base-T, FDDI, or ATM. That is my particular bias and experience. In fact, ATM as a data communications transport is not really available at this time and its congestion problems and broadcasts needs have not been resolved. As the reader is probably well aware of the fast transmission protocols hyped by vendors, magazines, and trade publications, I do not repeat the hype in this book. I do show how to analyze transmission channels and determine which is suitable for a faster solution and when it is the appropriate solution for a bandwidth bottleneck. The enterprise network is an integrated environment, not just an independent collection of segments, subnets, servers, hosts, and routers.

In fact, enterprise networking can no longer be viewed as merely a data connection, but rather should be seen as a fundamental infrastructure for all organizational and production activities. Business has become dispersed, internationalized, global, and mobile. Products and services are now less tangible than pizza, beer, automobiles, and other such "widgets" of economists' gross product modeling. Products and services have become commodities differentiated only by special delivery methods, price, informational content, accuracy and timeliness of the information, and secondary services themselves supported by informational content. Differentiation is dependent upon enterprise infrastructure.

Acknowledgments to readers and contributors

Enterprise Network Performance Optimization is all new material. Any overlap is specifically enumerated, and you are referred to my prior optimization and tuning books. Although optimizing the enterprise network overlaps LAN tuning, and performance of hosts and servers over the enterprise overlaps computer system tuning, the material is not repeated. The earlier books cover the specifics of using protocol analyzers, modeling performance, understanding traffic statistics, using CPU loading benchmarks and measurement tools, and matching RAM, CPU, I/O, caching, disk throughout, platforms, and system configuration with performance requirements. This book does not rise above specifics on coding, cabling, connections, statistics, and the configurations necessary for tuning operating systems; there are an extraordinary number of intentional details. Although some very detailed tuning techniques seem pertinent only to NetWare or NFS, extrapolate their applicability to other operating systems, media, or protocols.

I would like to thank the many readers of *LAN Performance Optimization* and *Computer Performance Optimization.* I would particularly like to thank those who wrote to elaborate on their solutions to performance problems, to question particular techniques for pinpointing bandwidth bottlenecks, to discuss the manageability and performance aspects of architectures and protocols, to show how some internetworking solutions were partial or represented poor integration, to show me better ways to solve computer integration and data communications performance problems, and vendors who provided product demonstrations that solve pertinent enterprise network performance bottlenecks. Special thanks to the many individuals and vendors who provided traffic capture and modeling tools, some of which are provided as slide shows or working demonstrations on the CD-ROM. Those ideas, issues, and tools fueled this book.

Contributors are listed in the appendix. Write to McGraw-Hill (sales and editorial loves to see letters) and the letters eventually get to me. Send E-mail, which is faster than turtle mail, via CompuServe 70370,1350 (which I prefer to Internet's npi@shadow.net).

Chapter
1

Overview

Introduction

This book is your guide to improving and tuning enterprise network performance. It is particularly appropriate if you design, plan, build, manage the resources, or maintain mixed platform or multisite data communications networks—or if you are troubled by the adverse performance of the enterprise network. *Enterprise Network Performance Optimization* shows you how to benchmark, analyze, and tune complex data communications networks. The reader should have a basic understanding of network technology. Although this book shows you the practical nuts and bolts about how to tune enterprise networks, it also provides significant insight into the mechanical, technical, and statistical operations of large data processing and communications environments.

Data communications networks are like spiders' webs. They typically begin as small structures with a simple purpose, but given success, they grow larger and more complex with larger nets and more infrastructure, as Figure 1.1 illustrates. Similarly, most organizations typically outgrow the initial LAN or host-based infrastructure and create consequential performance, integration, reliability, stability, security, and management problems. Furthermore, the interconnection of multiple networks, the integration or interoperation of different communication protocols, and just the geographical reality of the metropolitan, campus, wide area, or global communications typically create performance bottlenecks—virtually by complexity alone. Networks (much like spiders' webs) ensnare the unwary, collapse under their own excessive complexity, or fail catastrophically when overloaded, unbalanced, or extended past capacity.

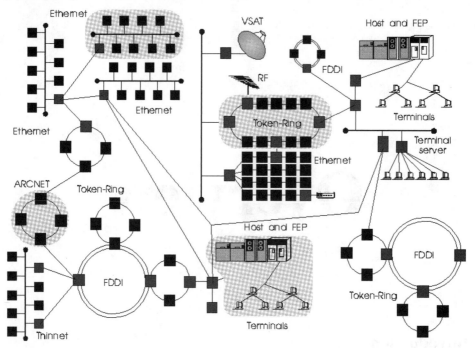

Figure 1.1 A metaphor of growth from original stand-alone computing facilities to the far-flung web of the enterprise.

Enterprise Network Performance Optimization addresses the crucial performance issues such as bandwidth, latency, blocking, and throughput. Key topics include packet and cell conversion, tunneling, encapsulation, translation, routing, network design, network architecture, protocol substitution, and workload balancing and reduction. Key techniques presented in this book include network design, performance planning, benchmarking, capture of traffic data, performance analysis, performance modeling (emulation, and discrete, state, and queueing simulation), and performance tuning. Other topical issues presented are economies-of-scale, platform heterogeneity, boundary performance, microsegmentation, virtualization, and routing versus bridging and switching. The book is a training manual, a "tips and tricks" look at complex network and data processing environments, an investigation and review of high-performance and high-bandwidth technologies, a troubleshooting guide, and an on-line multimedia hypertext reference.

This book is for all those seeking to improve computer application performance and data communication results. *Enterprise Network Performance Optimization* comprehensively explains how to reconfigure or rebuild large network infrastructures to run faster and provide better results. It helps you plan the information processing downsizing and build LANs and internetworks sufficient to support the transitional and long-term capacity. It helps you detect performance bottlenecks, analyze root causes of these obstructions, determine the best action, and then implement effective enterprise network performance optimization.

The MPC-compliant reference CD-ROM has 150 MB of material specifically oriented to the enterprise network. This includes a hypertext on-line reference file with the table of contents, text, full-color illustrations, index, keyword searching capabilities, and hot links to related topics, products, demonstrations, and definitions. The CD-ROM also contains multimedia sound recordings and video clips, published software benchmarks, product samples, demonstrations, and other tools. All the acronyms and definitions can be found in the glossary on the CD-ROM. These items are integrated into the text, referenced, and are included for a defined purpose. The CD-ROM also includes information sources that are indispensable for analyzing and tuning the performance of client/server, host-LAN connections, distributed computing, and data communications networks.

This book is about expanding the performance of networks to keep pace with growth, technical change, business information processing transitions, and data transmission requirements. It is about making networks perform beyond the norm—or to your organization's expectations. It is about getting the best return on investment for network infrastructure. It is about getting the fastest results. It is about getting the best quality results. It is about doing the most within integration, financial, time, or other resource limitations. It is about normalizing performance expectations and developing a practical methodology for enhancing performance. It is about tuning LANs, networks of LANs, host-linkages, the enterprise network, and overall operations performance.

As such, *Enterprise Network Performance Optimization* shows how to optimize the architecture, routing configuration, and the data communications environment, as well as the network hardware and software. This book is not about specifics of optimizing the performance of just one tool, just one operating environment, or simple heterogeneous networks. As soon as you optimize so specific a factor, a new revision is released, hardware changes incrementally, organizations create new mission-critical applications, and a new set of tips, tricks, traps, and help is suddenly available. Rather than just focus specifically, this book aims to define the underlying methods and detail the thought processes for effectively optimizing enterprise operations and making them go faster.

If you need information about planning, modeling, and tuning LAN workgroup operations, refer to *LAN Performance Optimization* (1993). If you require information about improving client/server, distributed computing or database application performance, applications servers, operating systems, and workstations (including the ubiquitous PC), refer to *Computer Performance Optimization* (1994). Both McGraw-Hill books are by the author.

Optimization is more than simply going fast. Going fast nowhere, going fast blindly, or choosing the wrong route is unproductive. Sometimes, people choose dead-end paths and lose sight of the real mission. This book retains the image of the real mission—that of providing a working, profitable, and viable data communications infrastructure throughout distributed organizations—and suggests ways to go faster with purpose. *Optimization* is the means to achieve your destination by the most economical route, with the least work and the best results. Solutions are presented with a mind to practicality and feasibility; organizations do not want Star Wars solutions for information age problems. Coverage includes computing platforms, network devices, system and network operating systems, application software, and utility software. The following list gives the performance targets covered:

- Transmission protocols
- Routers
- Gateways
- Hubs
- Network switches
- Out-of-band management tools
- Enterprise management tools
- Benchmark tools
- System and performance analysis tools
- Computer platforms
- Operating systems
- Graphical user interfaces
- Network operating systems
- Application software
- User activities
- Workload consolidation and reduction
- Network architecture
- Network configuration

While *Enterprise Network Performance Optimization* describes common techniques for optimizing performance, it also provides prolific examples specific to various network configurations, platforms, and operating systems. This comprehensive coverage is provided because multiple-network computing is the norm rather than the exception for most corporate and organizational data-processing activities. Computing platforms include proprietary network devices, host computers, Intel-compatible CPU-based computers (IBM-compatible personal computers), RISC-based engineering workstations, superservers, PowerPC machines, and Motorola-based Apple Macintoshes. Mainframe and mini-computer hosts are also represented, because most enterprise operations include integrated "big iron" computer centers. Operating systems include MS DOS (and PC DOS), OS/2, UNIX, AIX, SunOS, VMS, Macintosh System 6.x and 7, and PowerOpen, while Sun NFS, Novell NetWare, Banyan Vines, LAN Manager, LAN Server, NT Advanced Server, and PC LAN are representative of network operating systems, servers, and networked clients. This book includes abundant coverage of Microsoft Windows. Explore the on-line index or specify your own keywords for a hypertext search.

Additionally, *Enterprise Network Performance Optimization* discusses the relative merits of network architectures, collapsed networks, packet and cell switches, alternative transmission media and protocols, local data communications channels; network transmission protocols; and other special hardware that is championed by vendors to improve network performance. *Enterprise Network Performance Optimization* includes extensive coverage of hardware, software, host access, client/server and distributed procedures, utility software, and benchmarking tools. Utilities are represented by compression, balancing, and optimization tools, workload analyzers, tools that analyze network, system and software configurations and optimize them, backup machinery, and network analysis and management tools.

Purpose of book

The goal of *Enterprise Network Performance Optimization* is to show you how to squeeze more work throughput from the data communications and transmission environment. Primarily, this means increasing bandwidth, decreasing transmission times, improving processing speed with faster systems, and optimizing network configuration; secondarily, it means reviewing what you use the network for, how you use the network, and how you split workload and redirect work to secondary channels. It also means evaluating alternatives for improving results. Many tasks can be dramatically improved with better techniques, presentations, and applications. In other words, tasks may be optimized by finding better routes, avoiding misleading or dead-end paths, and matching network activity to the real, underlying mission and fundamental speed limitations. There are always better ways to get things done; sometimes we become inflexible and forget to realize, consider, or try other methods for tuning performance. In effect, *Enterprise Network Performance Optimization* delivers insight—some obvious, most practical—to improve productivity.

Although magazines have published many articles about tuning and performance optimization, the material is usually focused to a single operating environment and lacking breadth necessary for the enterprise network environment. Trade articles basically define a scattershot approach. Few books are available on this topic; most exist as specialized technical reference manuals from hardware or software vendors or third-party publishers. Furthermore, because there is no "typical" enterprise network, techniques for designing, planning, building, benchmarking, analyzing, and tuning extended networks must be more broadly-based. This book targets these gaps and provides information on benchmarks, analysis tools, performance tuning kits, and the methodology for optimizing extended network performance.

Intended audience

The intended audience includes all planners, designers, managers, and vendors of LAN, WAN, and enterprise network components, and anyone who uses an enterprise network and wants to tune operations and optimize network performance. Anyone who is intending to connect even one host or to extend the LAN to WAN sites or to mobile users will find this material valuable for its tricks and tips, new ways of looking at network performance problems, and realistic views about performance. This book is a useful teaching tool; it covers many systems and architectures from the vantage point of specific technical details but also with the perspective of problem-solving methods.

- Chief information officers
- Information planners
- MIS directors
- Industry consultants
- Network managers
- Data communications managers
- Vendors of connectivity products
- Vendors of high-end gateways
- Vendors of network switches
- Software developers building:
    Client/server networking applications
    E-mail networking applications
    Groupware (Such as Lotus Notes)
    Distributed networking applications
    Integrated enterprise applications
    Downsizing and transitional software
    Organizational "glue"
- Support technicians
- Help desk technicians
- Communications professionals
- Students of data communications

Because the possible return on investment of resources is very great when optimization is performed over a large base, the material in this book is clearly useful to managers in distributed network environments. Although the enterprise network typically represents mission-critical or profoundly fundamental operations for mobile, distributed, or world-wide data processing and communications, network managers are constrained by location, time, and the workability of solutions within integrated and heterogeneous network and computing environments. Nonetheless, network managers also tend to work within realistic budgets. Therefore, the replacement of perceived bottlenecks with the latest and fastest systems is strictly not feasible, both on a resource basis, but also on a functionality and feasibility basis as well. Many solutions are not cost-effective and are thus unworkable. The alternative is to locate the functional bottlenecks, then tune performance or find better ways to optimize results.

Book content

The content of *Enterprise Network Performance Optimization* is straightforward. This book defines performance issues, typical performance bottlenecks, addresses methods for quantifying and assessing performance, and suggests methods of improving performance. The book compares data communications networks with transportation networks and slow performance with the traffic planner's ultimate city nightmare, gridlock. Suggestions include general and persistent guidelines for optimizing performance that are applicable to most networks. There are copious references to recently published (hence, technically current) sources and trade articles should the reader require more information on a subject. Additionally, this book presents specific techniques, tools, and methods for tuning NetWare, Vines, SNA, HLDC and SLDC, DB2, NTAS, LAN Server, LAN Manager, PC Server, NFS, other network operating systems, other databases, and enterprise-based list distribution and routing services (such as ENS, STDA. X.400, and MHS). Because network performance is not just a function of network infrastructure performance, but also affected by the performance of attached platforms and operating systems, these are addressed as well. This includes UNIX, OS/2, MS Windows, database, client/server, and distributed or WAN-based processes. Moreover, *Enterprise Network Performance Optimization* goes beyond solving the obvious and immediate bottlenecks to review common difficulties. The book suggests reviewing the purposes of computing, assessing the mission as it should be, and trying to find shortcuts, better solutions, and new approaches. Suggestions range from performance tips to better commercial solutions.

Purpose for optimization

Optimization increases the financial value of the enterprise network. The investment is an involved installation of wiring infrastructure, WAN links, connectivity devices, computer equipment, expensive software, application development, workgroup procedures, and support staff. A dysfunctional network decreases productivity, slows or even halts production, and delays results. A slow network diminishes results—whether the results are the actual product, material for a new product release, organization newsletter, sales literature, customer support, shipping manifests and bills of lading, searches through on-line libraries or extensive databases, or an exploration through the electronic mail (E-mail) messages. A slow network translates into lowered productivity, additional labor expenses, and reduced profits. While computing capacity and disk storage facilities can be added in incremental units, the enterprise network transmission itself is a infrastructure not always so expandable. Therefore, it is wise to assess network performance. Discover if a network facility can be optimized before embarking on a potentially disruptive and uncertain path to microsegment, re-cable for faster protocols, replace connectivity devices with virtual switching, or upgrade protocols and network operating systems.

Just as additional mainframe capacity is expensive, additional enterprise network capacity is also (as it transcends single sites), and it is very disruptive as well. Incremental mainframe capacity can mean adding another mainframe. Incremental enterprise network capacity represents a more complex addition; it could require more PVC or SVC links, a shift from DS0 to ISDN or to frame relay, and new hardware to support the change in interconnectivity. It could also mean total replacement of a heterogeneous computing environment with one based on standardization. It does mean change.

Such change incurs major expenses. While incremental mainframe time is available from a time-sharing vendor, supplemental capacity for enterprise networks is not available from outside vendors. Incremental capacity may include permanent investments and changes in how the organization functions. Therefore, the customary goal of enterprise network performance optimization is to maximize bandwidth capacity and minimize the transmission latency.

As white-collar productivity is increased substantially for office workers—via telecommunications and widely dispersed networks—enterprise network performance optimization gains importance. The network becomes an inseparable cog in operations. Networked computers displace traditional data processing activities, such as data entry and retrieval. Therefore, the relative efficiency of networks translates into bottom-line profits. As an enterprise network gains influence and importance, effective optimization translates into sizable organizational benefits.

About the included reference CD-ROM

The reference CD-ROM incorporated with this book is MPC-compatible. The majority of the material requires MS Windows 3.x with some multimedia extensions. Although not absolutely required, sound and video highlights the presentation. The reference CD-ROM contains software, text, benchmarks, and other tools pertinent for network performance optimization. Most of the included programs are operating system-specific, mostly MS Windows. Refer to Chapter 6 for contents, file details, and installation requirements and instructions.

Structure of book

Chapter 1, this overview, presents the topics and the framework for *Enterprise Network Performance Optimization*. As an acknowledgment of the reader's limited time, Figure 1.2 shows the table of contents and flow of the book in a visual presentation.

Chapter 2 details a performance paradigm, that of a transportation infrastructure. Coverage includes the analogy of the city traffic gridlock. Definitions are given for the traffic measurement values of peaks and bursts, volume, transmission speeds, and delays. The chapter then presents the infrastructure for data communication wiring, LANs, MANs, WANs, and the workload contribution for various data communications operations.

Chapter 3 details typical enterprise network performance bottlenecks. Coverage includes network loads, transmission bandwidths, technical limitations, and configuration problems. It defines a bottleneck both in rigorous technical terminology and through practical examples. Less obvious enterprise network bottlenecks, such as complexity, network management, tunneling and encapsulation, user training, and reliability, are described here as well.

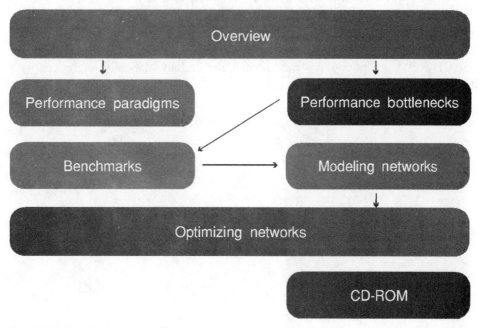

Figure 1.2 Organization of *Enterprise Network Performance Optimization.*

Chapter 4 presents the toolkit for enterprise performance analysis and optimization. This chapter explains benchmarks and defines a benchmark as an applicable measuring system based on applicable performance criteria. There is also an introduction to workload characterization as it pertains to the precision, utility, selection, and applicability of benchmarks as a performance analysis methodology. The chapter discusses tools for blueprinting, inventory control, network management, data capture, and performance optimization. Chapter 4 also demonstrates how to model network performance with spreadsheets, statistical analysis tools, benchmarks, iterative load application, network modeling tools, and disaster continuity planning tools, some of which are included on the CD-ROM. The emulation and simulation modeling tools include COMNET, Prophesy!, LANModel, BONeS, and GPS.

Chapter 5 is the primary focus of *Enterprise Network Performance Optimization* in that it demonstrates the methods for tuning network performance and reviews high-performance and high-bandwidth technologies. This chapter includes methods to enhance the process with different equipment and software, including unexpected substitutions and the reassessment of goals and realistic options for achieving them.

Chapter 6 describes the contents of the reference CD-ROM, the requirements for accessing it, how to install it, how to run the demonstrations, and how to apply the information to common computer performance bottlenecks.

The appendices contain acknowledgments to the people and organizations that helped with this book, supplied technical help and demonstrations, as well as numerous references to benchmarks, white papers, textbooks, other tuning guides, and tools. Additionally, *Enterprise Network Performance Optimization* contains a large glossary of acronyms and common terms; these are included on the CD-ROM in a popup hypertext format and are automatically referenced with hot links when they appear in the text.

2

Performance Paradigms

Introduction

You can throw FDDI at your slow Ethernet links, replace hubs with duplexed switches, and add other hardware based on common sense and experience. It may work as intended, or it may not. Until you see the traffic and its underlying patterns and actually simulate performance based on the patterns unique to your environment, you are likely to overstate tuning and optimization costs and even miss the target. You need to understand the traffic flow in a detail you never imagined in order to accurately, efficiently, and cost-effectively optimize the enterprise environment. This chapter compares the enterprise network to vehicle traffic and the network devices to different types of intersections.

An enterprise network consists of LANs, backbones, clusters of activity work-groups, hosts, gateways to those hosts, multiple transmission channels (some even parallel, redundant, or idled for emergency backup), and all the infrastructure necessary to maintain the network. It often is unclear when an enterprise network is a single entity or just a network of networks. You usually find out that the network is in fact singular when distant events create local repercussions, or when a local problem prevents remote users from performing their normal tasks.

A useful distinction between a network of networks and the enterprise network is that the enterprise network has multiple transmission paths between many sources and destinations; this implies a need to optimize routing paths. Also, the enterprise network may include basic LAN-to-LAN data communications connectivity overlaid with multiprotocol routing and phone, voice, facsimile, video, and other services multiplexed into the same transmission infrastructure. When the interrelationship is so dependent, you rarely can resolve problems at a local network level, because the problem is structural—or rather it is a problem with the enterprise network infrastructure. If you cannot address the bottleneck at a local level, you probably need to look at the big picture.

An enterprise network does not necessarily mean that a network is global, covers a country, or represents an entire organization. By literal definition, the enterprise network does service an enterprise, but that network must also include alternative point-to-point routes. It is insufficient to say a wide area communications infrastructure is equivalent to an enterprise network. In effect, the fundamental difference between optimizing the performance of local area networks and optimizing performance for an enterprise network is best summed up by:

- Speeding up city street traffic flow with a prudently positioned traffic officer
- Diverting traffic past the center city to avoid a series of slow intersections

There is no simple fix for the latter case. It is more than an issue of scope. An enterprise network is not a big LAN. The resources and requirements are not just scaled up. An enterprise network is even more than a series of LANs. While it is possible to view a single LAN as an efficient replacement for a sneakernet, the same is untrue for enterprise communication.

At a physical level, the geographic coverage usually precludes cost-effective or efficient courier delivery of information. Second, the integration of organizational activity probably transcends the network transmission channels; the network is not just a conduit but has enabled previously impossible or prohibitively difficult activities. Third, the enterprise network requires redundant channels for mission-critical activities and this presupposes routers, spanning trees, alternate transmission paths, and fall-back countermeasures in the event of route failure.

Transportation system traffic

A transportation system provides an excellent metaphor for the enterprise network. The metaphor is useful for envisioning its physical structure, the interrelationships between processes, locations, communication segments, and the logical design. Transportation systems are heterogeneous environments supporting different speeds, legal systems, rules, management methodologies, performance, and reliability factors. Furthermore, most people can visualize traffic and extrapolate it to the faster, more complex, and mostly invisible data communications channels.

The analogy for the enterprise network in *Enterprise Network Performance Optimization* is a transportation system. Just as people, goods, services, raw materials, and information (in letters) are delivered on planes, trains, cars, trucks, bicycles, feet, freighters, and other physical transportation devices, there are several methods and routes for delivering information. It is not so limited as to encompass just one LAN cable. Remember, the sneakernet and the faxnet *always* exist to supplement the on-line LAN. Furthermore, people can always deliver information in person and through the telephone. Sometimes gossip and private conversations are the preferred delivery conduit. Thus, logical data telecommunications and communications networks present options for delivery of information. The enterprise network increases these options.

The following eight illustrations show different types of traffic congestion. Although the channel appears to be the same street, do not be tricked into thinking that the street is the same; the street is not always unidirectional. Also, the causes for the traffic jams are not the same in each image. There are subtleties in each situation that create different types of performance bottlenecks—just as there are subtleties in enterprise performance bottlenecks. It is important that you can identify the conceptual differences between slow intersections, blockages, and managerial faults from bandwidth limitation as that pertains to networks.

The congestion begins for different reasons at different places, but often creates the same result—the traffic jam. If you can see how the problem begins or recognize that the traffic patterns are different, you are closer to your goal of improving enterprise network performance. The first illustration, Figure 2.1, shows the apparent network performance bottleneck caused by overload in the channel. Too many vehicles not only clog the street and degrade throughput, but because traffic backs up and creates a *queue*, it causes secondary delays at intersections that affect the other pathways.

Traffic dynamics

The CD-ROM contains a movie showing a day-long sequence of traffic congestion. The assumption that traffic is predictable, static, and uniform oversimplifies the dynamics of any traffic system. Anyone who has upgraded with expensive global fixes, backplane routers, high-speed switches, and other Star Wars solutions *and failed* to improve fundamental network performance will admit to the complexity of the traffic dynamics and functional dependencies. Although we think of traffic as a steady-state gridlock or slowly moving ooze, and in fact *traffic* is synonomous with the *traffic jam*, it really begins as a statistical trickle, growing in volume until it overloads the infrastructure. Even when the volume stresses the infrastructure and does not exceed its carrying capacity, peaks, bursts, intersection failures, or traffic accidents are able to overwhelm any system. The logistical dynamics are themselves as complex as the traffic flows. Although traffic dynamics create bottlenecks, the solutions are undeniably as varied as the causes.

This movie and the figures in this chapter declare that a forceful attack on network traffic bottlenecks is careless, expensive, and ill-conceived. As this book shows, instead, monitor traffic flows for arrival times, payload size distributions, error conditions, second-order effects, and the patterns for the bottlenecks which necessarily occur at intersections. Learn to recognize the subtleties. Recognize the patterns. Generate action plans for resolving overloads, accidents, bandwidth limitations, misrouting, poor traffic flow, and conflicting traffic protocols.

Figure 2.1 Traffic overload gridlock at an intersection, which might correspond to a network bridge, router, switch, or gateway.

Figure 2.2 presents a typical traffic gridlock as commuters are apt to experience at rush hours in that while one directional pathway is clogged, the other route is mostly clear and moving freely. A directional traffic gridlock may reverse directions between the morning and afternoon rush hours. Although this is not usually such a *visible* problem on the enterprise network, it is actually a more likely performance issue than the channel overload.

Figure 2.2 Traffic round-trip bottleneck caused by unidirectional congestion that might correspond to an overloaded server or an ISDN connection with one failed data channel.

For example, most LAN and client/server processes are based on messaging; there is the initial request, a message acknowledgment, and the actual service fulfillment. Typically, traffic is a two-way process. Commuting is a round-trip, albeit separated by about 8 hours. While you might consider the Federal Express delivery of a package a one-way process, consider how either Federal Express or the sender was notified to send a package in the first place. Also, damaged or lost goods probably create a considerable number of phone calls between recipient, sender, Federal Express, and Federal Express insurance claims.

Similarly, the data communication process is at least a two-way event. An inbound message or request initiates an outbound acknowledgment, some processing that in turn yields outbound traffic, and then inbound acknowledgments of receipt of outbound traffic. Figure 2.3 shows the traffic flow on a network. This flow diagram is simplified and does not show any intermediate hops, router spoofing, or sequence control, management overhead, or other traffic not directly related to the inbound message. However, flow does become more complicated on the multi-segment enterprise, and overhead messaging increases greatly.

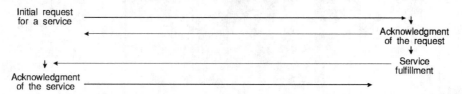

Figure 2.3 Schematic of network traffic flow.

The initial message is outbound from the source to the destination, while both the acknowledgment and fulfillment are inbound to the source. For example, an SQL request from a client outbound to a server could overload the server and create delay in fulfillment of the SQL request, but first it would cause a delay in responding to the initial message itself.

Directional overload in any direction will delay fulfillment of the round-trip process. While it is unlikely that the communication channels themselves are bi-directional—Ethernet, FDDI, and Token-Ring are certainly not unless they explicitly include a duplex hub and NICs—other network components are by virtue of their buffers, multiprocessing capabilities, and gridlock on other segments of the enterprise network. For example, inbound traffic may metaphorically back up on a network segment because the server NIC buffers are full, the server cache is busy swapping memory to fulfill earlier requests, or the server I/O channel is busy with disk activity. Although the outbound transmission channel is probably empty, there is the inbound bottleneck because the server cannot process all the incoming requests. As a secondary result, the network bottleneck is apt to spread and affect the outbound channels and create gridlock at channel intersections. This is just as the one-way traffic jam is very likely to cause subsequent slow-down on the side streets—which in turn is likely to jam up the traffic in other currently clear directions as well.

Figure 2.4 illustrates traffic prioritization, a technique used to exclude certain vehicles from the channel or provide enhanced service for particular classes of users. In this case, private vehicles are excluded from the street, and buses are provided with high speed lanes. This is typically employed in metropolitan areas to discourage traffic and reduce overall volume, reduce latency, and increase throughput by increasing the efficiency of the transport medium. By the way, you can liken data compression and run-length encoding to packing more people into buses (and other vehicles).

Network traffic prioritization is important for SNA and SDLC when spoofing is not employed to overcome message acknowledgment limitations, and particularly for real-time processes that are timing-sensitive. This includes video, synchronized sound and video, and physical control of apparatus. If physical devices are controlled by commands issued over a network with delayed performance, the data, the action, and the results tend to become unsynchronized with catastrophic results as users try to overcompensate for delayed feedback.

Figure 2.4 Prioritization with exclusive bus lanes (channels) much like DSLw or SNA router delivery prioritization.

Many protocols actually have the facility to implement traffic prioritization at the MAC-layer. For example, IBM Token-Ring protocol does include packet prioritization; yet most network operating system software does not implement it. Likewise, Ethernet in effect overrides transmissions with collision alarms. These alarms could provide useful information for traffic prioritization, but it too is ignored. By and large, little sophisticated software is available to take advantage of the burst patterns or statistical quiet times on Ethernet to improve throughput. On the other hand, FDDI-II, routers, and store-and-forward devices do provide some prioritization. However, when not intuitively implemented, it moves bottlenecks and performance problems to other locations on the network. "Traffic sheer" is a situation that occurs when slow-moving vehicles obstruct lane changing or slow devices limit throughput. Typically, you see this where the different classes of traffic and traffic moving at different speeds interfere with each other. This potentially occurs on the enterprise network above moderate traffic loads at all intermediate (interconnectivity) nodes. The net result is that traffic is constrained to the lowest common speed. Figure 2.5 shows gridlock, a problem when the traffic protocol is not rigorously adhered to or enforced.

Figure 2.5 Traffic gridlock at an intersection (intermediate node) due to secondary route backup caused when other problems spread, as might happen when a connection fails and creates a cascading network panic.

If you drive in a city, you know about gridlock—where drivers hog the intersection at a red light or block access to or egress from a rotary. However, data communication follows the protocols almost exclusively. The few protocol failures are caused by transmitter jabber, device failures, or channel defects. It is not as though a packet on the enterprise network is going to decide for itself that it will violate red lights, violate speed limits, and drive on the sidewalks; in fact most intermediate nodes will remove or isolate other nodes from the network if they cause protocol violations.

More likely, gridlock occurs on the enterprise network when any device fails to provide service at the requested (and required) speed and creates a backlog, bottleneck, or service failure. Technically, traffic arrival time exceeds service time in this case, and creates the queue. Switches, bridges, routers, and gateways are prime examples of intermediate nodes that can back up because they cannot maintain wire speed. "High speed" switches can fail when multidirectional arrival times exceed the switching time, or when overlapping requests for a channel cannot be fulfilled because the first request blocks subsequent ones until it is fulfilled. This is not channel overload, but congestion at an intersection.

Furthermore, when multiple routes are established for redundancy or bandwidth expansion, the failure of secondary routes can overload the primary ones and create service gridlock somewhere else on the network. Many people misdiagnose intersection bottlenecks as bandwidth congestion simply because there seems to be so much traffic on a particular segment. Check traffic for source and content (with a protocol analyzer or with a management agent).

Bandwidth congestion can also be caused by chronic or temporary capacity reduction. While slow workstations, servers, and other primary nodes are likely to spawn an intersection bottleneck as in the prior figure, *segment* bandwidth congestion is typical for internetworks, WANs, and campus or enterprise networks with mixed-speed transmission channels. Nor is this the same problem as previously shown in Figure 2.1. The infrastructure bandwidth is not insufficient; rather there is a channel transmission chokepoint within the infrastructure. Examples include the on-demand modem line activated to replace a failed T-1, a failure of a secondary route or spanning tree, the delayed enabling of the spanning tree secondary route while a router converges, or a T-1 line used as a conduit between two FDDI LANs. Another example is represented by the channel temporary restriction due to construction, as illustrated in Figure 2.6.

Bandwidth congestion is also likely where protocols are encapsulated or translated—not because of the process time—but because of the differences in the packet sizes, slot time frequencies, and data payloads. With translation or encapsulation process in real-time (or with a minor latency), the arrival and process time differences swamp the intermediate channel. For example, the switch over from one bus to another itself is not so time-consuming as the problem that occurs when a full bus disgorges its passengers to a bus with a smaller capacity or less frequent service times. That creates a backlog, not really a bandwidth limitation, but rather a service failure.

It should not be too surprising that lumbering traffic creates more conflict than smaller vehicles. However, the difference between buses and cars is not as large as the difference between large and small network packets. Recall that network traffic always moves at wire speed, and gets to that speed instantly. Small packets *will* provide better performance on LANs supporting early token release (FDDI and Token-Ring). As graphs in *LAN Performance Optimization* show, larger packets are technically more efficient, but most network traffic consists of short messages. These effects, while marginal for most LANs, are unlikely to have anything but a detrimental effect on the enterprise network. If you study the illustration in Figure 2.7, the true effects of larger vehicles is not that they obstruct and delay traffic flow, but rather that they simply take up more of the road—available bandwidth in the terms of the data communications network.

Figure 2.6 Traffic congestion due to a temporary channel capacity reduction (as is typical with LAN-to-LAN connections over a shared or multiplexed WAN link).

Large vehicles may require more processing or resources at primary nodes such as servers, workstations, or overflow buffers. Although it does take longer for the larger buses to change lanes or make turns than for smaller cars, performance statistics show that routers maintain throughput with larger packets by virtue of their relative efficiency; however, larger packets typically create higher rates of packet losses than small packets because of the buffer overrun.[1]

Figure 2.7 Traffic congestion due to traffic composition. In real enterprise networks, traffic mix has a profound effect on overall traffic load and end-to-end transmission time.

Figure 2.8 shows that surprises do occur. The bicyclist wants to get where he is going quickly and without regard for others; the truck is either lost and going the wrong way or trying to back up to a loading dock at a building. Similarly, the enterprise network has its own surprises, whether it be mechanical breakdowns, flaws, defective packets, lost packets endlessly repeated and forever circulating, or loads you really do not want competing with the normal traffic. A typical unwanted load on the enterprise network is in-band management traffic.

---

[1]*Review of Router Performance*, Tolly, Kevin, Data Communications, July 1993.

Figure 2.8 Gridlock due to traffic obstructions (bicyclist in street and truck going wrong way), as happens when switches or routers forward defective packets or packets to the wrong subnets.

Figure 2.9 may seem just like the obstruction in Figure 2.6 or the surprise in Figure 2.8, but it actually differs. The police officer moving the barricade and waving the cars forward is actually supposed to improve and facilitate network performance. However, as is typical, the confusion and fear over a watchdog creates a slowdown. Similarly on the enterprise network, the management protocols—SNMP, routing table updates, and remote (RMON) management traffic—can create a performance bottleneck by being too aggressive and too demanding of network bandwidth. For example, the SNMP MIBs define over a hundred separate statistics, which would cause the enterprise to choke if these were continuously monitored for every node and transmitted in real-time to a central monitoring station. There must be a balance between exercising the bandwidth for overhead and control, and transmitting the actual traffic load.

Figure 2.9 Gridlock due to device or process overload (police arguing with traffic). This happens as routers converge on alternative routes. The problem is not lost bandwidth per se, but that it is temporarily inaccessible.

Obviously, fax and voice conversations are not always the accurate and efficient replacement for electronic communications. However, when you visualize and review enterprise networking, the network is not a service wrapper, but rather an integral infrastructure. If you can separate the traffic and processes from the infrastructure, you do not have an enterprise network; rather, you just have a conduit without a real relationship with the traffic. If you remove the traffic from a city, it ceases to exist as a social infrastructure. Similarly, an enterprise network exists only for the infrastructure required for the entire organization.

Enterprise network infrastructure

Consider the demographics of daily New York City commuters, where 1.7 million people drive themselves, car pools move another 250,000, buses and ferries transport about 500,000, 4 million walk (wholly or as part of the commute), the subway moves 2.9 million intramurally, and trains carry about 500,000 (composite of the Long Island Railroad and ConRail). Past railroad strikes create more havoc than the 250,000 people directly inconvenienced; the effects are to double or triple overall commuting time. The bottleneck effect was even worse when an interstate bridge over the Mianus River 40 miles west in Connecticut collapsed in 1988. The impact was more than that of a link failure; it was that of infrastructure crisis.

Even though alternatives to the railroad existed and side roads bypassed the Mianus bridge failure, the bottleneck affected the system more than peripherally. While some might insist that network channel or device failures are more analogous to train strikes or roadway failure, the daily traffic bottleneck is endemic to New York and other cities and cannot be merely solved with wider highways, faster trains, priority traffic lanes, and longer airport runways. Those techniques create the usual result that when some commuters shift to other forms of transportation, new commuters fill the void. The effect is status quo.

The material and passengers brought by airplane still need to be delivered by car, truck, rail, or bus. Motorists still must oblige traffic lights and deal with errant streams of jay-walking pedestrians, double-parked vehicles, parking automobiles, and the delay at parking garages—and still probably walk a block or two to the elevator in their office building. The transportation system is not just the highway channel, but all the interchanges, destinations, and method transfers. Similarly, an enterprise network is not just the transmission channel, but all the possible alternate channels, bridges, routers, packet translators, processors, storage disks, and slow software in the infrastructure. If you resolve one bottleneck with a wider path—as with an extra lane on the interstate highway—you are as likely to displace workload elsewhere, slightly increase overall throughput, but not fundamentally address network performance.

On the other hand, the 1987 savings and loan debacle gutted Wall Street of financiers, brokers, traders, lawyers, and support staff, and *that* indeed improved the traffic flow. If you want to improve your enterprise network infrastructure, just consider how you can remove workload from the system. However, just as the displaced workers were dissatisfied with this stock market "correction," you can be certain displaced workers will be equally dissatisfied with workload reductions. The key point is that you can improve enterprise network performance either by reducing the workload, compressing the traffic, offloading the workload out-of-band (such as by telecommuting), by improving the performance of some to the detriment of others, or by addressing the data transportation infrastructure in its entirety.

Network enterprise aspects

The traffic, management, reliability, and performance issues of the enterprise network infrastructure are described by topics presented in subsequent sections:

- Channel capacity
- Bandwidth
- Signal speed
- Route
- Protocol overhead
- Delay (latency)
- Accidental latency
- Peaks and bursts
- Frame size
- Volume
- Lost traffic

Although these topics seem unusual, they are pertinent to the statistical analysis, design, management, and optimization of the enterprise network. They are relevant to the construction of a useful paradigm for modeling the structure, interrelationship, and operations of the enterprise network. Technically, traffic models are based on arrival and service distributions, which in turn define loads, latencies, protocols, routes, and errors. This presentation has significance for modeling a complex network in Chapter 4.

Channel capacity

Obviously, the most important influence on highway traffic is the overall traffic-carrying capacity. Capacity is actually quite complex in that it is a function of the number of parallel lanes, the traffic speed, interlane traffic shear and interrelationships, and delays that occur on transition points such as entrance ramps, toll booths, interchanges, curves, and design reconfigurations. Individual drivers and their quirks have profound and tragic effects, too. Singular traffic events, such as accidents, breakdowns, or inattentive drivers can affect the channel capacity as well. Similarly, these metaphors parallel performance characteristics in data communication networks.

Bandwidth

Posted highway speed is equivalent to the data transmission signal speed, the wire speed, or the channel bandwidth; they are all equivalent. This is usually either zero or the protocol transmission speed. For example, modem speeds are 9600 bits/s, Token-Ring is typically 4 or 16 Mbits/s, Ethernet is 10 or 100 Mbits/s, and ATM ranges upward from 25 Mbits/s. When Ethernet supporting a client/server load, for example, transports only 232 Kbits/s as a maximum throughput, this is a function of the entire infrastructure including the servers, NICs, and workstations, not the channel bandwidth itself. Even Ethernet maximum speeds of 1.8 to 3.2 Mbits/s, which are characteristic of wireless transmission, would have extra capacity for that client/server load. Synchronization requirements in parallel lanes, as with dual NIC configurations in servers, Fibre Channel, and protocol messages are comparable to traffic shear and traffic correlation. Chapter 3 puts these numbers into graphical perspective; bandwidths and channel capacities vary widely, and leased line speeds and a router backplane do not even fit on the same graph because there is such a magnitude of difference.

Signal speed

The channel signal speed is the bandwidth or the wire speed. These terms are effectively the same. The signal propagation speed is the speed at which the signal traverses the channel, and is mostly a factor of the medium and the speed of light. There really are no intermediate speeds, as might occur when highway traffic creeps past an accident.

Route

The path by which information travels from source to destination is rarely an issue for LANs; there is only one delivery path. This is true even for FDDI (where the secondary channel is a counter-rotating transmission path activated when the primary connection fails); it is usually not an overflow or extra lane. Even when hosts are connected to workstations through a single bridge or gateway, the route is usually simple—you have just one choice—through that bridge or gateway. Multiple LANs, WANs, and hosts connected together with routers usually only represent a singular point-to-point between any source and any destination. When the network connects an enterprise and comprises multiple paths, multiple bridges, routers, multiport hubs, spanning trees, parallel channels, and switchable links, the issue of best route becomes important. The aspect that separates the network of networks (with a binary hierarchy) from the enterprise network is that the enterprise network does require optimal data traffic routing for optimal performance. There are many routing protocols for enterprise networking with different performance and integration effects; these are compared in Chapter 5.

However, it is important to recognize that when there are route options, routing decisions become important. The route may be as simple as choosing a modem connection at the error-free speed of 2400 baud rather than a noisy and unreliable 14,400 baud connection with V.32bis compression that slows to compensate for line noise. The route may be as complex as factoring multiple channel speeds over connected segments, device buffering rates and latencies, blocking times, and the need to avoid saturated links. Just as a truck driver may trade slow speeds, intersections, and accidents in an obviously shorter center city route for the longer city circulator, you may optimize enterprise network paths based on speeds, loads, delays, and hops.

Protocol overhead

We usually measure highway traffic in terms of number of vehicles. However, data communication vehicles are more complex. Data and messages are encapsulated inside cells, frames, or packets. This is akin to measuring people transported or tons of cargo moved *within* the vehicles on the highway. There is an overhead associated with this packaging. For example, ATM cells are 53 bytes, of which 5 are overhead, and data payload comprises 48 bytes. The payload is also likely to include other overhead as well when route and path information, configuration and process control, TCP/IP messages, and SNMP are transported as data in cells. The difference between the effective and the actual data load is part of the protocol management overhead used for synchronization.

Additionally, most protocols impose other overheads. Just as highways have innate loads—signal lights, toll booths, slow curves, climbing lanes, construction zones, and traffic that just manages and protects the highway (snow plows, street cleaners, police vehicles, and highway cleanup and repair crews)—which subtract from overall highway capacity, data communication channels impose management overheads. Keep-alives, failure indicators, polling, and status markers subtract from overall communication channel capacity. Although pneumatic counters on highways count vehicles (or axles as a proxy) *out-of-band* (sideband) to the traffic, most network overhead and network management is carried as part of the traffic itself and is called *in-band* management. Sometimes, even so-called side-band signaling is merged into the data stream as a multiplexed signal; as such, it is bandwidth utilization. This is particularly acute on the enterprise network because of the inevitable use of in-band signaling for networks alerts, control, and monitoring.

Imagine tracking highway traffic and accidents with messages delivered by people in vehicles who have caused, witnessed, or think they have seen a traffic tie-up! Such a process would be absurdly slow if the messengers could even get through or bypass the event. Imagine reporting a developing problem—a driver drinking intoxicating beverages—using the highway at highway speeds before something does happen! The message would arrive too late.

Delay (latency)

When you are stuck in a traffic jam on a highway, you wish you could only go the speed limit. Delays are formed in at least three ways. First, the data communications signal propagation speed is finite, a significant portion of the speed of light at 50 to 90 percent. The latency station-to-station for VSAT is about a quarter of a second, cross country it is about 100 ms, transglobal it is about an eighth of a second, and nodes communicating on opposite ends of an Ethernet will experience 9.6 µs—several orders of magnitude less. Remember the speed of light as a functional latency. It may seem small, but distributed applications typically incorporate several round trips of transmission delay per transaction. It may seem minor, but it does accumulate.

Second, transitions force you to reduce transmission speed. Just as trucks slow at curves, intersections, or ramps, data payloads are switched, routed, repeated, encapsulated, translated, and buffered on data networks. Because the enterprise network is a compound LAN, MAN, and/or WAN, there are apt to be multiple transitions and inherent latencies at each transitional device between channels and segments. Traffic that spans multiple segments is delayed not only by the inherent signal speeds, but also by interconnecting network devices. Bridges, routers, gateways, and switches are intersections with delays that range from 40 ms to several hundred or thousand ms. These delays are at least as significant as the delays caused by channel signal propagation speed (1.6 ms for FDDI, 25 ms for an average Token-Ring, 51.2 ms minimum after an Ethernet collision).

Third, just as accidents, construction, and slowdowns cause delays on highways, these defects and backups create significant delays on the enterprise network. Bad cells and packets force retransmission from the *initial* source. Signal loss, signal errors, CRC errors, bad data, data out of sequence, device or channel failures, device or channel overloads, and other events are analogous to traffic accidents. Slowdowns cause packets to overflow buffers, and because networks cannot allow traffic to crawl along in a stop-and-go fashion, these are typically dropped with the expectation of a later retransmission.

Accidental latency

The ultimate network delay is an accident or incident. If a link fails, traffic on either side of that pipe is halted, or at least detoured through longer or slower links on the enterprise network. Frame relay lines fail, ISDN connections drop, and modem connections wither with line noise. Network devices can saturate a line with gibberish, or fail to relay or route data traffic at all. Workstations, servers, and hosts fail outright or experience processing problems that create performance backlogs and stoppages. Sometimes you see a cascading shutdown on the infrastructure. Even when linkages are supported with alternate or backup pipes, there are likely to be significant delays while these routes are switched on-line, enabled, and routers, gateways, and switches are updated with new physical and logical routing information.

Device failures or slowdowns create backlogs just as events on the highway create traffic jams and gridlock. If a server slows down, client requests can saturate the channel and prevent completion of ongoing tasks or fulfillment of subsequent tasks. While this is typical on LANs, a cascading failure is even more disastrous on networks of LANs and the enterprise network. The cascade is primarily caused by network devices not recognizing that there are service failures, by network workstation software not realizing network service slowdowns, and users, who are increasingly pressed to complete tasks, rerequesting services. SNA time-outs, failed status, and lack of response create this gridlock. Note that data communication gridlock is not like highway gridlock in that the transmission channel clogs with stationary vehicles. Instead, you may observe "normal" traffic levels and bandwidth utilization, that is until you parse the content of the traffic and recognize that time-outs, IP acknowledgments, duplicate messages, router address table updates, and server and hosts service broadcasts are flooding the network. Some people call this cascading failure a *network panic*.

A network panic has systemic ramifications on the enterprise network that adversely affects all users and processes. Furthermore, the cause of the traffic jam is quickly lost as the bottleneck spreads from the initial point of impact, much as a pebble dropping causes ripples on a pond, as Figure 2.10 illustrates.

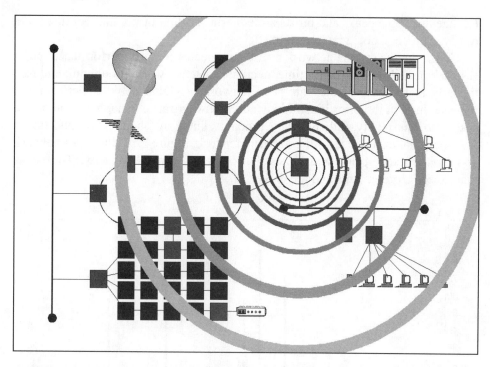

Figure 2.10 Network panic spreads nearly instantly throughout the enterprise network and collapses the network and services infrastructure.

Deciphering the cause for the panic is not as easy as flying over the infrastructure. The tools for enterprise network management are primitive. There is no comparable big picture tool like the helicopter, which is a true out-of-band tool. Ping and pulse tell the logical status alone and do not provide timing and qualitative information about how well the device is running. For example, a router may be available but not actually functioning while corrupted routing tables are being rebuilt with new network information (which isn't forthcoming). Protocol analyzers are merely counters, albeit with some more advanced capabilities with protocol decoding. However, more advanced tools (such as NetView, OpenView, and UniCenter) try to provide status and qualitative information so that you can monitor host, server, router, gateway, and channel performance. Yet, none of these tools can resolve the panic after-the-fact in a multiprotocol environment. Often, the only solution is a phased shutdown and restart of the network segments, clients, services, and connectivity devices.

Peaks and bursts

The paradigm of peaks and bursts is that an extreme cyclical or recurring workload will exceed the capacity of network components, thus creating a momentary bottleneck, or worse, a lingering traffic jam or network panic. In extreme cases, these peaks and bursts collapse bridges, routers, LAN segments, and processors.

Realize that these peaks and bursts are either a function of normal workload or a runaway process, not "gratuitous" traffic.

Generally, the sudden network traffic jam creates sluggish performance, process backlogs, slowed response time, decrease in actual work throughput, and particularly in the case of the enterprise network, a global network performance malaise. In fact, the prior illustrations and highway traffic metaphor should clue you in to this reality, although you probably know all about peaks and bursts from personal experience with your own network. Figure 2.11 illustrates network and typical traffic peaks and bursts that overload channel capacity. The peaks that exceed a preset threshold are darker.

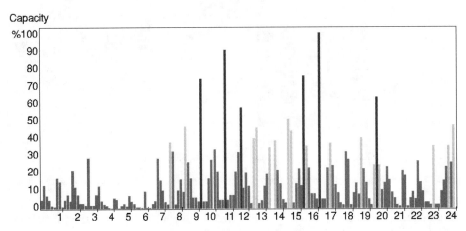

Figure 2.11 Peaks and bursts in network traffic are indicated by the darker colors .

However, there really are no peaks and bursts of network traffic different from any other network traffic; it is all a matter of workload. In fact, if you raise the bandwidth ceiling on the graph in the prior figure, the problem of the peaks seems to vanish. Note that limited bandwidth can delay and postpone interactive or batch processes; when addressed, however, the same workload may create peaks at different times in the day as the peak load hits the network earlier and is completed sooner in the day, as Figure 2.12 shows. The increased availability shifts peak load to 11 a.m.

Many network designers and managers believe that increased channel bandwidth solves these bottlenecks, and so they improve cabling plants and LAN server infrastructure to support faster signaling protocols, switched media, or greater parallel bandwidth (as with duplex connections). Alas, that is not the entire solution. Consider what happens if you lower the bandwidth ceiling in the prior figures so that the patterns are more obvious. Do you notice how previously hidden peaks and bursts create new bottlenecks? Sure, you have the original bottlenecks—in spades—but you still have the same pattern of peaks and bursts. Figure 2.13 shows channel loading with the transmission capacity reduced. This should imply that if you do increase channel capacity, the inherent burstiness in the traffic may create similar load bottlenecks elsewhere in the system at the

current traffic levels, or similar bottlenecks with only modest traffic load increases. Traffic always exhibit this pattern of peaks and bursts at all load levels.

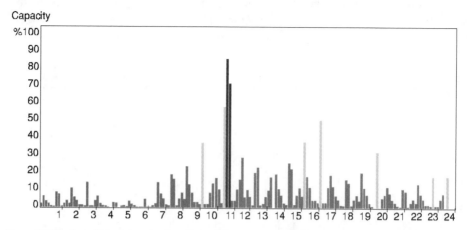

Capacity

Figure 2.12 Peaks and bursts in network traffic are indicated by the darker colors, and are made to seem irrelevant by increasing the bandwidth or channel capacity. In this example, the bandwidth is at least doubled with either parallel channels, duplex transmission, or switched media infrastructure upgrades. The 11 a.m. spike indicates that a previously-constrained process now has the headroom to complete.

By the way, this is a trick you will want to add to your arsenal of tools. Before assuming that a particular solution will work, stress-test the effectiveness of that solution by making the situation more acute and observing the results. Reverse the proposed solution and evaluate the results. This is much cheaper and less politically damaging than trying a costly solution that fails.[2] It is usually useful but not always logically accurate. In this example, reversing the bandwidth in-crease demonstrates that the traffic is inherently bursty regardless of the scale.

The research on data communication loads basically shows that there is a morning rush hour or singular afternoon commute on the enterprise network only when the aggregate load exceeds the carrying capacity of *some parts* of the infra-structure.[3] Most network traffic has self-similar characteristics of peaks and bursts, which, when aggregated, create the reality of a traffic jam. More likely, a singularity creates the illusion or appearance of a peak- or burst-based overload. This includes a freak network accident, a faulty device, or a process that is a bandwidth or service hog. However, it alone is the failure point and you will need to address the singularity appropriately.

Similarly, aggregated processes do not create peaks and bursts by themselves; it is a statistical event that you perceive the aggregated traffic loads to have peaks

---

[2]This stress-testing may falsely assume conversity or reversibility. Also realize logically that the converses, inverses, and negations of true statements do not necessarily yield true statements.

[3]*On the Self-Similar Nature of Ethernet Traffic,* Will Leland, Walter Willinger, Murad Taqqu, Daniel Wilson, ACM SIGCOMM 1993, *The SIGCOMM Quarterly Publication,* Volume 23, Number 4, October 1993.

and bursts; in fact, all network traffic at any level will exhibit peaks and bursts. It is the nature of LAN, client/server, and heterogeneous data communications traffic. Bottom line: peaks and bursts are statistically aggregated workloads that overload parts of the network infrastructure. They are a normal part of LAN-type traffic, and best characterized by fractal mathematics of self-similarity.

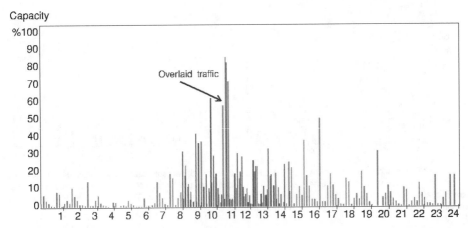

Figure 2.13 Peaks and bursts in network traffic loading are intrinsic to data communications traffic regardless of the channel capacity or transmission infrastructure. One time frame is overlaid on another to show the self-similarity.

Enterprise network traffic is never uniform; traffic has complex and intertwined patterns. It can be functionally decomposed to individual traffic streams, each with its own characteristics, dependencies, and internal patterns. Figure 2.14 illustrates these traffic streams, but note that there are still peaks and bursts into each colored stream.

Network traffic is a quasi-random mix of cells, packets, frames, datagrams, prioritized streams, and a few other message formats. It is not random because most processes require a message acknowledgment, as with TCP/IP, or requests initiate responses, subsequent data transfers, and activity fulfillment. Although there may not be a direct correlation between any two messages, there are underlying patterns to the network traffic. Burst mode windows, which allow transfer of multiple packets with a single burst confirmation message, does not make the network any more random; it merely changes the stream characteristics.

If traffic patterns were uniform, you could calculate loads by multiplying the size of the messages by the rate of messaging. Note that this does not work well. It may provide a primitive extrapolation of network load where resources limitations are not active constraints. Also, linear estimates do not scale to larger networks. For example, if 20 clients create 1000 units of network traffic or 10 percent loading (of available bandwidth or server capacity), 40 identical clients will not realistically or necessarily represent 2000 units of network traffic or 20 percent loading. The characteristics of the traffic streams, the protocols in use, and the data processing methods will determine actual network loads. They are most likely to be higher, but could be even lower when the extra load is represented by duplicate work, cached data, records locked and inaccessible, or economies-of-scale often experienced in client/server transaction processing. Benchmarks based on a linear loading algorithm (such as PERFORM2, PERFORM3, and NETBENCH) are not very representative of the workload on your enterprise network environment, but will demonstrate approximately at what absolute scale peak traffic levels will saturate the enterprise infrastructure.

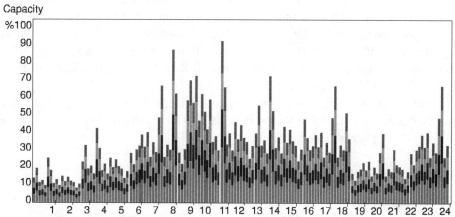

Figure 2.14 Network traffic streams.

Frame size

Frame size has a profound effect on LAN and WAN performance, and an equal or even more pronounced effect on the enterprise network. Typically you worry about the larger packets as they have a greater tendency to overrun buffers in intermediate nodes, the more perverse effect for the enterprise network. The processing effort for larger packets is the same as for smaller packets, so larger packets are more efficient. However, router queues are more likely to fill with the large packets. On the other hand, given two networks with the same travel bandwidth utilization, the one with more packets (that are obviously smaller) will tend to have greater Ethernet collision rates or longer token rotation times. Chapter 4 shows the simulation of different packet sizes. Additionally, translations or encapsulations among different protocols create additional traffic as there is usually something greater than a one-for-one transition. For example, some packets from Ethernet may represent 60 ATM cells after translation, and while ATM bandwidth is (15 times) greater than the Ethernet the difference does not necessarily offset the increased latencies with the cell streams.

Volume

Peak traffic levels that cause bottlenecks really represent a problem of scale relative to available capacity, rather than a problem of the peaks or the bursts themselves. This is a traffic volume issue. Aggregate traffic that exhibits peaks or bursts that overload enterprise network resources really indicate a traffic volume overload that happens to occur during the peaks or bursts. This is a function of capacity and load at all levels of the network infrastructure. Traffic volume is really a two-sided issue, that of load and that of capacity. Specifically, the load must be less than the capacity. You can solve either side of the equation; either reduce the load, or increase the capacity. If volume overloads are specifically related to peaks and bursts, you can attempt to modify those temporary loads. However, you are more likely to find that the traffic overload is a basic problem.

The issues that arise for balancing this equation for the enterprise network include the scale itself for the traffic volume, the sensitivity of the peaks relative to that scale, and the uniformity of the data transmission and processing infrastructure. The network is hardly ever uniform. Recall that if you alter the transmission bandwidth capacity, you are very likely to create performance pressures elsewhere or at different times. This is akin to widening a highway to create center-city traffic jams a few minutes later as the increased traffic tries to locate parking on-street or in parking garages.

Data communication highway

The metaphor of the enterprise network as a transportation system is pertinent because we can divide the infrastructure into highway segments with different routes, and devices, services, and processes with different performance characteristics. This is useful for two reasons. First, most people can envision and understand the traffic jam on the highway. Second, there exist some traffic planning tools and methods for city planning, but most city and traffic planning is performed ad hoc also because resources are rarely allocated for it. A considerable amount of planning is not feasible or practical within the political environment. The same can be said for data communications networks, however, the enterprise network is absolutely vital to commerce. Furthermore, we are actively paid to improve enterprise network performance, so we will find practical solutions.

You may want to extend the paradigm of the data communications network to include transportation systems other than just highways and streets. Consider enhancing the metaphor to include airplanes, trains, ships, couriers, and other physical transportation methods as suggested previously, but scale it up for the volumes typical in networking (of hundreds to thousands of packets/s).

We measure highway traffic with helicopters, police, traffic counters, traffic modeling tools, demographic information, and drivers with mobile phones. At two levels, comparable comprehensive tools do not exist for the enterprise network. First, most of the highway traffic measuring tools are out-of-band. For example, the helicopter flies above the traffic grid and the traffic counters are read by hand or by telephone. While you might make the argument that drivers who are stuck in traffic and make a cellular phone call to a radio station to convey highway status are part of the channel and the traffic, they still represent out-of-band measurement. Although they are stuck in traffic they are not using the highway for transmitting status. Second, enterprise network traffic is amorphous, encapsulating, and orders of magnitude more complex and faster than highway traffic. As a result, the highway traffic metaphor is appropriate for the enterprise network when scaled for volume, transmission speed, and other differences:

- Enterprise network transmission speed is greater.
- Enterprise network transmission volume is larger.
- Data content and function is relevant to data traffic analysis.
- Data highways are multifunctional.
- Data highways carry greater service diversity.
- Data highways are more configurable.
- Data highways change more frequently.
- Network integration is more complex than transportation.

Mechanical infrastructures

The enterprise network depends on many components you ordinarily would not consider as part of a network or even realize are fundamental to computer operations—until they are unavailable or create maintenance and performance problems. Just as highways have infrastructure, so does the enterprise network. While you are likely to think about infrastructure in terms of physical components, such as communication wiring, bridges, routers, switches, modems, and network interface cards, instead look at the logical structure of processes, sources, service providers, and users. The technology of networking is such that physical proximity is often immaterial. From a performance aspect, routes, speeds, intersections, termination processing time, and round-trip times are probably more important. Performance analysis requires you to create an abstraction of the network; this is necessary for any benchmarking and modeling, as in Chapter 4. Typically, you create an iconic representation of the enterprise network, as shown in Figure 2.15. Notice that you can define equipment and outline its usage on such a diagram; some people create process flow diagrams.

However, you still need a physical blueprint of your network as this picture does not show you where the wiring actually goes. You will probably need an architectural blueprint showing precise positions of conduits, vertical risers, entry points, and manholes. Tools such as SysDraw and netViz (which are profiled in this book and included as presentations on the CD-ROM) provide other approaches to creating and maintaining full diagrams of wire routing, wiring closets, color-coding, and devices. For the purposes of this book, the following three illustrations define the network; these pictures will form the basis for benchmarking and modeling throughout *Enterprise Network Performance Optimization*. They are also useful to show you what tools are available for documenting and analyzing your network. Chapter 4 describes several tools in some detail. Figure 2.16 shows a global view of an archetypical organization's enterprise network, while Figure 2.17 details North America. Figure 2.18 shows the main campus in San Fernando Valley.

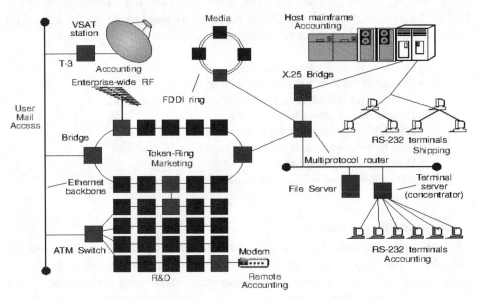

Figure 2.15 A simple logical schematic of an enterprise network.

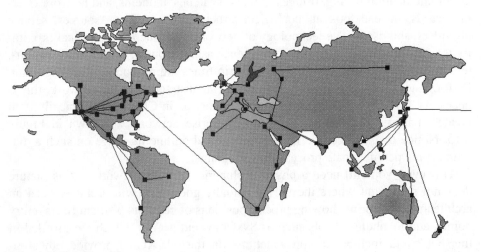

Figure 2.16 Global view of an enterprise network.

Another artifact of the enterprise network is broadcast traffic and packets that circulate in search of their destination. This is akin to a lost rider on a subway or the driver that goes around the cloverleaf looking for the missed exit, doubles back over the route searching for the destination, or wanders aimlessly in lieu of any destination. Lost traffic does not generally exist on LANs as packets are not repeated by source stations (FDDI and Token-Ring), or are terminated at network endpoints (Ethernet). Lost traffic is an issue for enterprise networks because it gets repeated or routed between multiple subnets and segments.

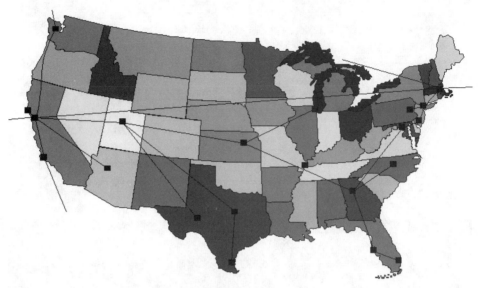

Figure 2.17 View of the North American component of an enterprise network.

However, the complexity of enterprise networking often creates leakage and losses of data anyway. They typically occur at network junctions and intersections. Just as traffic accidents usually occur at intersections, the same problems occur on networks at the intermediate nodes. Bridging, routing, switching, and gateway devices cannot retransmit data if the device itself is overloaded *or* the pathway on the other side is overloaded. Routers create black holes where something escapes. For example, you cannot progress from an off-ramp that has a traffic bottleneck on it any more than a router can transmit to a network in nearly constant collision or exhibiting slow token rotation times. As a result, network problems can only be inferred by lack of error messages or acknowledgements.

Leaks and losses

Data leakages and losses refer to the misdelivery of information or incompletion of delivery, rather than to theft or breach of security. Data leakage and losses are actually rare in the scheme of networking because most processes, particularly SNA, SDLC, TCP/IP, IPX/SPX, and similar internetworking protocols, require a one-for-one acknowledgment of every request.

That is, the transmission window size is 1. Therefore, any request that does not receive a response, will send out a duplicate request. If the request has still not been fulfilled, the operating system on the requesting device will ping the server for status or keep-alive notification. If the messaging protocol requirements are not fulfilled, the application or connectivity device will allow a session or connection to time-out.

Figure 2.18 Campus portion of an enterprise network with physical wiring paths exposed.

Also, PBX-like switches typically have no means to buffer inbound data when a circuit has already been switched between communicating stations. Although switching times are very short—anywhere between 32 and 340 µs—messages directed to a file server could exceed that rate. Another type of leakage occurs when routing tables are defective. This can and does occur with routers, gateways, servers, and any other devices, including switches that map physical locations to the logical network addresses. Figure 2.19 illustrates traffic leaks and losses that occur at routers and gateways.

Pressure and volumes

Traffic is measured as vehicles per unit time. This is as true for highways as for data communication networks. Network vehicles are bits, cells, frames, or packets. Network traffic-level work throughput are usually measured in bits/s or frames/s. Unfortunately, you can only measure actual work accomplished, not the level of work that you desire the network—traffic or data—to move. You also cannot physically measure process completion time delays. Figure 2.20 shows the difference between the pressure and measurable volume.

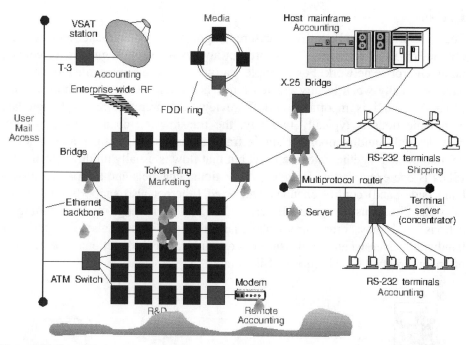

Figure 2.19 Network leaks and losses occur at routers, bridges, gateways, and switches. These intermediate network nodes can become traffic "black holes" due to configuration errors, overloads, insufficient arriving-packet buffering, and bottlenecks ahead of them that gridlock them. The traffic loss (indicated by a lack of signal acknowledgments) spawns more network traffic replacement, which merely adds more pressure to the congestion.

When the enterprise infrastructure is overloaded, the throughput does not indicate the source pressure from users, processes, and devices wanting to apply more workload to the network. Because network transmission protocols dictate signal speed and transmission rates, you cannot *measure* a backlog pressure or queue length as you can at a toll booth or off-ramp. Servers and on-line tasks (with exception of batch queues and the more sophisticated operating systems) do not provide queue information as is traditionally maintained for mainframe batch process queues. With NetWare, Vines, or NTAS, the information is primitive at best. However, you can estimate or *model* the pressure, queueing delay, or latency with modeling tools for any traffic protocol and network environment, as shown in Chapter 4.

Figure 2.20 The difference between a volume and the pressure creating that volume.

LANs

The enterprise network traffic pressures begin on LAN segments and become obvious on LAN segments. These are not always the same segments, of course, on an enterprise network. For example, an overflow at a remote site can create a backlog locally when locally created requests for remote service cannot be fulfilled. As such, it is important to briefly review LAN architectures. Although data communications is typically two-way, the transmission channel itself usually provides only unidirectional or single frame transmission. Frames and data can be multiplexed together into one flow, but that flow is usually unidirectional also. ISDN is one example that supports bidirectional channels and a separate out-of-band (side-band) control channel. Duplexed Ethernet also supports dual transmissions directions. However, Ethernet, FDDI, Token-Ring, modem communications, and most other protocols share the media and also compete for bandwidth. Even when the pressure to communicate is bidirectional, the LAN is often unidirectional, as Figure 2.21 illustrates.

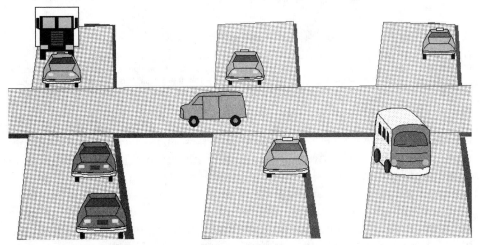

Figure 2.21 A LAN is a unidirectional street when traffic tries to merge and enter.

Even when wiring hubs, MAUs, CAUs, repeating hubs, and other wiring concentrators physically enable a star-wired infrastructure, the LAN is still logically wired as bus, ring, or star configuration with all the limitations inherent in the basic communication protocol. The channel is still unidirectional, but the entrances to the channel are not as direct, as shown by Figure 2.22.

Intermediate nodes are bidirectional interconnections to unidirectional segments or subnets. Duplex signaling provides two-way traffic by establishing a dedicated concurrent inbound and outbound channel (possible by disabling the collision on Ethernet). This technology creates private LANs through microsegmentation and is useful when the extra bandwidth can provide more headroom.

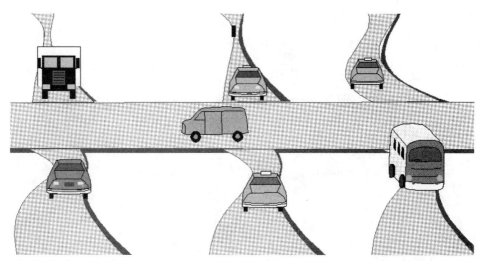

Figure 2.22 The hub or wiring concentrator is also logically a unidirectional street.

The *segmented backplane*, as sold by SynOptics, ODS, Cabletron, Retix, and other vendors increases performance through microsegmentation. In other words, the LAN bandwidth is increased by creating more logical channels where existing stations are redistributed to separate channels. Traffic is routed between segments or channels when necessary. In other words, you have created two or more highways with a limited intersection, each highway designed to serve a different area. This improves performance if and only if loads can be partitioned such that inter-LAN traffic is minimized. This technique is well-documented in *LAN Performance Optimization*. However, the jargon of backbones, backplanes, switching hubs, and backplane routers confuses the technology somewhat.

Kalpana, Artel, Synernectics (3Com), and others have created a different LAN-based routing architecture that improves performance by partially relieving the bottleneck of the shared media LAN. This switchable backplane device is typically not an internetwork router, but rather a device for LAN connectivity and centralized wiring. The fundamental difference is that media bandwidth limitations are overridden by constructing a backplane with enormous bandwidth (200 Mbits/s to 5 Gbits/s) and buffering incoming packets into a rotary-like process. The LAN is widened to accommodate multiple streams.

Although the actual connections from the stations to the router are established by the media- and protocol-limited bandwidths, the shared channel on the backplane can provide ten or several hundred times the throughput of the hub or backplane wiring hub. The downside is that the latency of such backplane routers due to the extra buffering and processing usually exceeds 150 μs. When multiple backplane routers are employed for a distributed internetwork—certainly the minimum definition for the enterprise network—the one-way latency usually exceeds 0.5s for one hop, 1.5s for two hops, and 3s or more for multiple hops. This technology is shown in Figure 2.23.

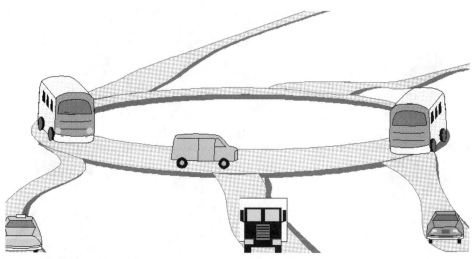

Figure 2.23 The router in a backplane is a rotary which directs multiple vehicles concurrently.

The newest LAN technology, the packet switch, also has applicability to enterprise networking. The switch creates a physical connection between any two stations for a period just long enough to transmit the cell, frame, or packet. This physical connection provides full media and protocol bandwidth. The media is not shared. The bandwidth is not allocated among all the stations wired into the switch. Essentially, this architecture bypasses the two-dimensional architecture of the LAN by creating three-dimensional bypasses and overpasses. Figure 2.24 shows a multipass metaphor for this switching bypass (cut-through) technology.

Switching technology was initially envisioned for ATM-based wide area internetworking. However, this technology has applicability for all LAN protocols because it can dedicate full bandwidth between paired devices on the network. There are two disadvantages to switches. They block duplicate connections to sources and destinations during an already established connection by virtue of the virtual paths (although you can overcome this by installing multiple NICs in key nodes or fronting the nodes with routers). To date, the switching times require about 200 to 980 μs—the same or slower than routers, although that should fall to about 25 μs before 1996, according to SynOptics product designers. Blocking occurs because a switched connection is a virtual connection between only those two points. When one of those points includes a shared device, such as a router, server, or gateway, other devices are excluded from communication access for the duration of that connection.

A switch is a modernization of the old-fashioned operator patch panel. A switch typically connects two stations together for voice, fax, or data. The illustrative switch is the PBX, the central piece of telephony as found in most organizations. A switched circuit is a virtualized connection, that is a logical and intermittent physical connection, between two (or more) devices.

In the case of LAN and enterprise networking, the switch provides the full bandwidth available with the protocol and the media to those attached devices. It is *virtual* by virtue of the temporary and exclusionary nature of the connection. However, because it is exclusionary, all other devices are prevented from reaching the connected devices during that connection. The best example of the switch is the drawbridge that prevents cars from crossing while ships pass underneath. The other limitation is that the connection is temporary and exclusionary, thus priority transmissions can be prevented from gaining immediate access. The best example is the ambulance or fire engine stuck behind an opened drawbridge. Also, there is a switching time while the drawbridge closes and gates reopen to allow the traffic to once again cross. This is true for LAN switches, too.

Figure 2.24 The switch is like a highway overpass that bypasses the shared media. This typically also prevents mid-circuit switching. The bus almost hidden on the lower overpass is committed to its path and cannot be rerouted or redirected with a jump to another circuit.

As shown in Figure 2.25, a drawbridge provides the switched right-of-way to traffic on one dimensional plane through the intersection, and offers absolutely no chance of passage for blocked transmissions. See Chapter 5 for a discussion of switching and the performance effects of cut-through or store-and-forward switching. The confusion for blocked broadcast messages is left to Chapter 4; it is a serious congestion control problem for switching technology and a fundamental shortcoming of LAN protocols routed to the switched-access enterprise network.

Figure 2.25 The switch is also like a drawbridge that cuts through a shared media environment and blocks access for the other traffic. Multiple connections are possible, but not between the same physical points at the same time.

Manageable operations

Unless LANs are managed well or managed centrally for the benefit of enterprise network users, no amount of effort can solve enterprise network performance problems. Just as you cannot expect traffic to move freely from major highways into a congested city, you cannot expect enterprise network clients and users to have efficient and timely access to congested LANs, hosts, and other remote services. You cannot control the metropolitan area traffic until you can control and manage the infrastructure as a whole. It is a political and functional necessity that LANs and global services attaching to the greater enterprise not create chaos and spread unnecessary traffic. If you recall, the enterprise network is not merely a collection of internetted resources; rather it is the infrastructure that often includes LANs. You will be unable to optimize enterprise network performance unless you can see and control the big picture.

Firewalls

A *firewall* is any mechanism to protect network stations, subnetworks, and channels from complete failure caused by a single point. This is particularly important in the enterprise scenario because you will not want any network segment or subnet creating a traffic jam that causes gridlock or a widespread network panic. The firewall itself is usually a network device—and an enterprise network actually requires a firewall at each gate—which is typically a bridge, router, or gateway. It is designed to prevent unauthorized access by hackers, crackers, vandals, and even prevent your own employees from accessing privileges internal information (such as sales accounts, human resource information, and network configurations). The firewall is like a castle moat surrounding the enterprise.

No amount of firewalls, security, access control, and other typical LAN management techniques will control network performance unless you can control all the components. This means—and Chapter 5 will present these concepts in greater detail—that traffic must be managed on the segment and subnet level so that segments and subnets perform well, traffic between segments must be contained within the internet bandwidth, and processes that transcend the enterprise network must be built to perform within the functional internet latency and bandwidth limitations.

MANs

A metropolitan area network either represents a campus network and a scaled up internetted LAN, or it represents a scaled-down wide area network. Although some have seen FDDI (with its one magnitude improvement in usable bandwidth over Token-Ring and Ethernet) as the ideal connective backbone, it only represents a small increase in potential throughput. It is not the ultimate solution to MAN requirements. Also, SMDS, T-1, T-3, and similar high-speed PVC services are not the ultimate solution for LAN connectivity either. Not even SMDS—at 44.7 Mbits/s—provides more than half the performance of FDDI. (Several figures in Chapter 3 illustrate the performance differences between common transmission protocols and transmission services.) As organizations push performance to the desktop, increased bandwidth will also be required for backbone traffic interconnectivity.

Load and latency limitations

The problem is that LAN traffic has peaks and bursts, and when piped into a backbone, easily overwhelms the intermediate nodes and sometimes even the backbone channel itself. Consider the traffic scenario when wrecking crews clear the highway of damaged vehicles and the roadway of debris after a collision. Any bottlenecks do not clear quickly as drivers vie for a lane, accelerate to speed, and maneuver around less-rapidly accelerating vehicles. It takes a while—sometimes through the commuter rush hours—until the primary and secondary effects of the tie-up finally clear.

The same is true for MANs after a blocked or overloaded channel clears. The backbone is a two-way transmission path typically routed over a single unidirectional channel. Even an FDDI backbone is still a unidirectional channel, albeit that FDDI supports early token release for simultaneous sequenced multiple packet transmissions. Although data frames accelerate instantly to the signal transmission speed, the clients, servers, hosts, and network nodes, like rubber-necking drivers, do not always recover so quickly from a problem. The buffers and stacks need resetting, and the clients and servers must funnel requests to the CPUs and disks. Ultimately, the backbone access protocol is first-come, first-served. Routers and gateways, which are supposed to provide priority queueing and filtering, cannot keep up with the sudden bursts, and overload secondarily.

WANs

When you break apart the enterprise network infrastructure as it is today, wide area networks are not very different from LANs or MANs. The difference between the MAN and the WAN linkages is usually in terms of distance, available channel speed and the larger traffic burdens on MANs. As you have seen, distance has a relatively minor effect on the overall communications latency because of the high signal propagation speed. Bandwidth is usually more of an issue. Although there are and have been high-speed WAN linkages available, such as SONET and VSAT, they are expensive. Typically, while WANs represent traffic problems because the long-distance channels are limited in bandwidth, more likely, they represent a management and control problem.

Let's face it. How do you manage and control operations in another city at a distance? Reach Out and Remote Control remote control software is part of the solution. But how do you solve a hardware switch setting or software configuration problem at a distance? The answer really is the same as a city traffic control system which can be regulated from a central location based on traffic loads, latencies, and information provided by video cameras and traffic patrols. You centralize the expertise and send out on-site support when you have no choice.

Bandwidth limitations

However, the initial focus of WANs was different from on-demand networking in that they were designed with low speed links (9.6 Kbits/s) supporting SNA for remote transaction processing. SNA is very compact, was designed within the limitations of slow dial-up channels, has a priority support mechanism even with high levels of TPS, and has sophisticated congestion control initially built into the protocol and added to the primary features of the host front-end processor. Contrast that with LAN congestion control features: collisions, packets dropped, increasing token rotation time, time-outs, and verbose retries. Client/server and the typical routed LAN load is bursty. Chapter 5 addresses bandwidth-on-demand solutions, but also the more easily implemented solutions which include regulating WAN traffic, installing firewalls, and assessing the true performance issues of distributed processing and replicated databases.

Costs

What can anybody say about highway infrastructure costs? Put the problem to a vote. Swing a bond referendum. Literally, show the importance of maintaining a state-of-the-art infrastructure for business continuity, competitiveness, product development, and efficiency. Calculate return on investment (ROI) for infrastructure projects even they are dispersed and enormous, or raise competitive pressures and the losses from failure to maintain forward-looking enterprise information services. While these seem like for-profit business concerns, no organization, even governmental service organizations, are beyond such issues.

Infrastructure is very expensive. It is much like a bridge foundation with the bridge or the bridge ramps. It has limited return until the entire structure is complete and productive. If you analyze requirements based on isolated workgroup operations rather than as an enterprise activity, you are likely to suffer the performance problems of the fiefdom-managed interconnected LANs—you will never be able to address the problem because you never have control, never see the big picture, or can never control the network integration and configuration issues. Network bottlenecks have a way of escalating into organizational ones.

Shared media vs. dedicated bandwidth

The newest technology for LANs and the enterprise network is the packet switch. It supersedes shared media. In the case of a typical LAN, such as Ethernet, Token-Ring, or FDDI, the available bandwidth and wire speed is shared among all the devices on the LAN. The shared media protocol determines how the bandwidth is distributed to the network devices.

Bandwidth on Ethernet is shared on a first-come, first-served basis, but this is modified by collisions; collisions cause devices to retry after a random exponential multiple of 9.6 μs. Devices rarely get 10 Mbits/s for more than a packet or two; the rest of the second is shared among other devices. Divide the node count into the bandwidth to estimate shared resources available for each node:

Available bandwidth = 10 Mbits/s / number of nodes

(However, that arithmetic is very inexact at best.) Similarly, the token passing algorithm allocates the bandwidth on Token-Ring and FDDI, and typically no station gets more than the THT (that represents the maximum frame or packet size) every TRT. In other words, each device shares approximately:

Available bandwidth = THT / TRT

Multicasts and broadcasts are more efficient in the shared media environment because a single packet can provide information content to many or all the nodes on the network. Most NOSs have timer packets or synchronization packets that network nodes use to reset internal clocks. On dedicated segments, broadcasts and multicasts are replicated by cloning the cast packets for each dedicated link.

Workload contribution

Every vehicle on the highway contributes to the traffic. Even vehicles in the breakdown lane cause onlookers to slow and observe. Even traffic on seemingly unrelated segments affect vehicles on other segments by delaying switching at intersections. However, highway traffic is not duplicated and gratuitously propagated onto other segments, as it might be on the enterprise. Fundamentally, though, they are the same. Round trips require bandwidth, messengers and messages reduce the available bandwidth, and in-band management traffic has a direct correlation while out-of-band management (via commuter news radio) shifts loading in both direct and indirect ways.

Just as you assess highway performance on all channels, at all intersections, and end-services such as on-street parking, garages, and valets; so too you will want to rate enterprise network traffic on all pipes—LANs, PVCs, modem connections, and wireless or VSAT links—check flows at junctions, and qualify the performance of hosts, servers, global resources, network workstations, and terminals. No device on the enterprise network is so passive that it will not affect performance. Furthermore, no device on the enterprise network is so innocuous that it cannot create a bottleneck. Even connectors—which are inactive and inert, which do not create cells, frames, or packets of any kind, and which seem functionally insignificant—can cause, absorb, and induce noise. Stray noise can partially reflect back onto the enterprise network and this will be interpreted as bandwidth applications. It thus represents a real workload.

Workload characterization

Modeling LANs, components, and the enterprise as an entity is part of designing, planning, assessing, managing, and troubleshooting the enterprise network. Typically, you will need to generate an experimental workload to check network performance, reliability, functionality, or bandwidth. This is a critical job. Extrapolating LAN and host traffic to a proposed enterprise network or existing data processing patterns to remote enterprise network processing patterns is about as futile as extrapolating traffic from other cities to gauge the performance of New York City traffic. Any similarities are quickly diffused by the differences.

In fact, there is a considerable amount of research on typifying workloads for different telecommunications and data communications architectures. Some of it is referenced in *LAN Performance Optimization* and *Computer Performance Optimization*. Some are embodied by benchmarking tools such as PERFORM3,

NETBENCH, SPECbase, and other transaction processing-oriented tools. The short answer is *do not use them*, as Chapter 4 forcefully details.

They do not reflect reality in general, and do not reflect your enterprise circumstances in particular. Block file transfers do not reflect client/server processing. Even suites of user applications (such as the BAPCo benchmarks) do not really reflect how your organization functions. Even suites of tests that capture real data from a live network (such as the LANalyzer demo on the CD-ROM) do not reflect circumstances other than the particular time and case of that capture; it has no relationship to your environment whatsoever except by mere chance. However, these tools are useful for the very specific purpose of stress-testing channels, interconnections, and services to see how much more load they might be able to accommodate, or if it is possible to overload network channels or devices. The test is not likely to show you which device currently represents a critical path for adding more TPS, adding new services, or expanding the enterprise infrastructure.

In fact, programmers developing applications for deployment over the enterprise network should carefully design benchmarks and measurement systems *before* coding, so as to avoid the very typical problems of building client/server applications that work for one or two users but fail when deployed for the workgroup. You do not want to drive your application down your enterprise highway and shear off the vehicle roof on a bridge, as so many trucks with oversized loads do on low bridges on the interstate.

Test early and often. Test on the live enterprise network. Otherwise, simulate the signal propagation delays over WAN links, and latency through multiple bridges and routers; a test device, called a *data channel simulator*, imposes set delays on transmissions so that developers can simulate propagating a database to a remote site, or replicating entered data to distributed databases. The data channel simulator is useful for stress-testing, demonstrating performance of network equipment, and injecting channel impairments and bursts.

Also, calculate the expected loads, expected DP service latencies, client workstations service latencies with the actual deployed Intel 286 machines or X-Windows terminals (rather than the souped up development machines), and transmission latencies across the enterprise network at the fully deployed levels. Workload characteristics include those issues. You certainly do not want to develop and deploy an application that works well stand-alone but cannot work with 400 concurrent sessions, that will not scale, that will not tolerate 7s internetwork delays, or locks up nearly constantly with record or file lockouts.

Scalability

Enterprise network architecture promises that loads can be scaled almost infinitely, a promise that is about as true as improving highway traffic by widening the highway with more lanes. The extra channels alone do not relieve the critical path. Some critical paths are not a function of width, but rather of length. Traffic at speed cannot go any faster by adding more lanes; this is a latency issue. In other words, the mythical man-month is still a month in duration and you cannot affect the duration of the process with more resources. Some things do scale, some will not, and often the overhead for managing extra resources exceeds expected gains.

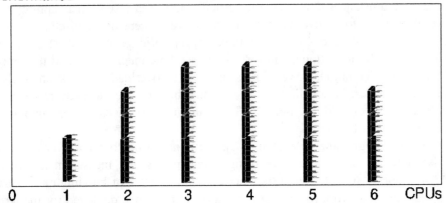

Figure 2.26 Adding more resources in parallel does not always provide greater throughput, as is the case with additional symmetrical multiprocessors.

Scalability is an issue of planning for deployment and workload growth and of process design, but also workload characterization. You can certainly add more parallel channels, install more symmetric multiprocessors, and replicate more database servers. However, consider the decreasing utility of additional resources within the enterprise environment. At some point—probably sooner than you think and hope— adding more resources serves no purpose and may even degrade performance (as Figure 2.26 shows with the addition of more CPUs to a platform supporting symmetric multiprocessors). The typical LAN processes and distributed LAN applications do not always scale over an enterprise network because of inherent time limitations in the file server and limitations in resource performance. Scalability is an issue that must be addressed as part of any enterprise network performance analysis and optimization.

Transaction processing

Transaction processing loads are different for different environments. SNA is very efficient in its native environment. The same is not true for the small packets of translated or encapsulated SNA data on LANs or on an enterprise network. Imagine a slow convoy of trucks on a multilane highway allocating an entire lane to themselves and discouraging other vehicles from accessing the other lanes through them. They jam up the highway at exits and entrances because vehicles cannot pass through them, as Figure 2.27 shows.

This happens when SNA is encapsulated or translated into LAN protocols for transmission on an enterprise network. It really is a question of load. When SNA or transactional load consumes too much bandwidth, it will create this routing nightmare. When employed at low levels merely to integrate essential but in-termittently-needed services into the enterprise network, service integration is cost-effective.

Figure 2.27 A convoy of buses can curtail access to the intersection for other traffic.

Unless the convoy is broken into smaller groups or allocated its own channel (such as the extreme lane from exits and entrances), the stream and the bursty LAN-type traffic cannot coexist. In fact, different vendors solve the coexistence problem in one of these two methods, as Chapter 5 elaborates. Some vendors try to prioritize the streams and the bursts with routing mechanisms; other vendors multiplex the traffic on separate T-1 channels and continuously reassess channel (and thus bandwidth) allocation based on the composition of each type of traffic. Multiprotocol prioritization works until the threshold of collaboration is sus-tained, after which only channel allocation demonstrably works.

Backplanes vs. hubs

A hub is a device that provides for centralized wiring connectivity. A hub is a powered MAU, an integrated signal repeater, and a means to concentrate data network wiring. A hub is an efficient method to comply with EIA/TIA 568 and EIA/TIA 569 premise wiring standards, consolidate the media, and centralize the wiring closet. Some hubs are *managed* in that you can switch out failed stations, gather statistics about traffic and error rates, or reconfigure the logical network without switch jumpers, cables, or stations. In the case of 10Base-T or newer 100Base-T, the hub replaces the bus coaxial cable usually necessary for Ethernet. A Token-Ring hub (such as the IBM CAU) may have options for extended wiring distances, logic circuits for testing continuity and correct wiring, and management for the LAN. An FDDI hub is really SAS concentrator, or a sophisticated extension cord with multiple outlets.

A backplane, on the other hand, provides a high-speed data communications channel with perhaps 10 or 100 times more bandwidth than the protocols it supports. A backplane is like a hub in that it provides wiring concentration and centralized management. However, it is a logical consolidation of the distributed LAN-wiring systems. The backplane also provides internetworking connectivity among multiple LAN segments. In effect, most backplanes are multiple configurable hubs which provide inter-LAN routing and LAN microsegmentation. Backplanes consolidate bridging, routing, and gateway services into a single chassis.

The distinction between hubs and backplanes is critical. A hub is simply a wiring abstraction, whereas the backplane represents the fundamental architecture for providing increased LAN, WAN, and enterprise network performance. At the simplest level, the backplane is a hub of hubs; however, it really is more because it isolates hubs into separate LANs and hosts into managed segments, and then connects those subnets into an internetwork with routing services.

Integration

No matter how you plan an enterprise network, integration is a major issue. If it isn't a primary design constraint—if you are building a network with all new components—it will be in a matter of months as newer technology must be fit into the preexisting framework. Effective use of the enterprise network requires a free flow of services and information, free in that it is available from any network station, but nonetheless secured in that secrets remain protected.

Transportation systems were not always so integrated either. It is something that has evolved from necessity. Trains could not share the tracks as the gauge varied. Early roads were not wide enough for truck traffic because they were designed for horse-drawn carriages. Ships were designed that required special berthing facilities. Today, standardized freight containers fit on truck trailers, train beds, ship elevators, crane gantries, and within special cargo airplanes. The transfer from one transportation method to another is efficient and integrated.

Although data encapsulation and translation is perceived as the means to unite different protocols, integration is not only an issue of packaging on the enterprise network. If it were, SNA and LAN traffic coexistence would not be such a challenge. Transmission speeds and latencies must mesh. Perhaps more importantly, the software guiding all these complex interactions from source to destination—the routing and the data processing—must fit together as well. As Chapter 5 explores, integration is very complex, made so not only by the number of choices in software, hardware, computing platforms, and network connectivity devices, but also by the complexity of enterprise network-enabled applications.

Conclusion

It is important to realize that the enterprise network is neither an interconnected LAN nor a replacement term for wide area networking. There also is a difference between a network of networks and the enterprise network. This difference is best shown by the enterprise network with multiple transmission paths (routes) or an integrated infrastructure that consolidates basic LAN-to-LAN data communications connectivity overlaid with multiprotocol routing and phone, voice, facsimile, video, and other services. As a result, you cannot resolve the performance problems at a local network level. Enterprise network performance problems are structural. Chapter 3 shows you how to look at the big picture, define the bottlenecks endemic to the enterprise network, and locate the critical paths.

Chapter

3

Bottlenecks

Introduction

This chapter describes enterprise network bottlenecks and intermediate devices, and how they increase latency and cause performance problems. This chapter is specific to enterprise networks and relates the traffic analogies from Chapter 2 to real events. Bottlenecks on the enterprise network usually evolve from a series of applications, significant components, and structures. Very rarely will an enterprise network have just one bottleneck or any singular identifiable defect. Individual LANs, a server, a router—these are slowed by a singular defect, but problems combine to create a traffic jam on the enterprise. Just as a traffic jam represents more than a single vehicle, the enterprise network will not be undermined by a single failure point. At least, it shouldn't.

Definition of bottleneck

Although this material can be found in *Computer Performance Optimization* in much greater depth, it bears summarizing here. It is worth looking at the bottleneck in both general terms and technical definition. In general, a bottleneck is a choke-point, the place where too much traffic tries to fit through too narrow a passage. Specifically, a *bottleneck* can be caused by any system component—hardware, software, subcomponent, process, task, or person—that delays the results beyond the expected completion time. Some synonyms for an enterprise network bottleneck include:

- Obstruction
- Obstacle
- Stoppage
- Impediment
- System strain
- Congestion
- Traffic jam
- Backup or backlog
- Resource limitation
- A slow or long route
- Interrupted pipeline
- Error

- Slow-down
- Overhead
- Barrier
- Overload
- Narrows (or narrowing)
- Link failure
- Gridlock
- Processing overrun
- Bad route
- Inactive channel
- Accident
- Excessive pressure

Although these terms are actually useful when you are staring at a protocol analyzer and wondering what to do, there is also a more formal description for the bottleneck. Formally, a bottleneck represents the critical path on a *performance evaluation and review technique* (PERT) called the *critical path method* (CPM). The bottleneck is the path without *slack time*. That is the path (or multiple paths) holding up the completion of the project or the path (or multiple paths) that delay later activities. The path lacking slack time is the limiting performance factor in the enterprise network, as Figure 3.1 illustrates.

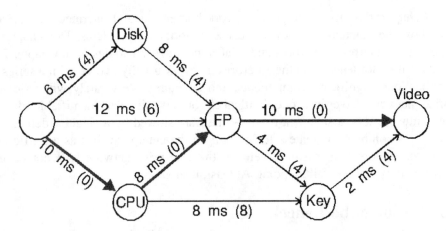

Figure 3.1 A critical path in CPM shows no slack time as indicated by the thicker path and zero slack time. Numbers in parentheses represent the amount of slack time.

The bottleneck is the critical path; it is usually not one device, such an overloaded router or a router with buggy software, but is the chain of all intermediate and terminal nodes on that path. Therefore, there are usually many active and in-force bottlenecks on the enterprise network at any time—the enterprise is a big place, after all. Some bottlenecks may preclude a LAN segment from operating efficiently. Some bottlenecks may be immaterial to your mission, and although critical to someone else in no way impede your mission. Adding slack time to the critical path—that is, speeding up that path or simplifying that task so that it completes sooner—generally shortens project completion times; when the critical path can be relocated with minor model adjustments, the system is very sensitive to tuning and perhaps not sufficiently stable for effective optimization. Improving performance on paths that are not so engaged—are not critical paths—does not affect performance, as the inconsequential performance improvements in Figure 3.2 show.

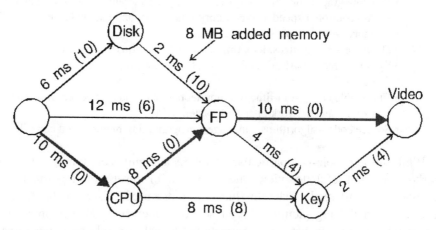

Figure 3.2 Adding slack to paths with slack time does not affect overall performance. Paths with slack time are narrow; those paths that constrain completion time are thicker.

Bottlenecks are limitations, restrictions, choke points, blockages, constrictions, impediments, slowdowns, application clutter, constraints, qualifications, and protocols. It is important to recognize that there are tradeoffs and compromises that often must be made between performance, speed, functionality, reality, time, and, of course, the availability of resources, political astuteness and organizational culture, financial or otherwise, to resolve the bottlenecks. Bottlenecks are not always obvious. What may masquerade as a performance problem may be in fact a training issue or a problem of misuse, mismanagement, poor configuration, or other similar flaws. Lack of qualified personnel, cost or difficulty in upgrading to a new software release or hardware platform, and the time required to retrain users for new operations and software can represent bottlenecks. System integration, manageability, sheer size or distribution, compatibility, or lack of documentation and support from a vendor can severely restrict performance. Specifically, the enterprise can be broken down into the following components:

- Infrastructure—design, architecture, protocols, implementation
- People—skill levels, training, support facilities, and experience
- Intermediate nodes—hubs, repeaters, bridges, routers, switches, and gateways
- Organization—politics, goals, locations, funding source, stability
- Operations—people, task complexity, time-criticality of task, and management
- Hosts—mainframes, minicomputers, and servers
- CPUs—processors, motherboard design, and operating systems
- Network—protocol, transmission speeds, hops, tuning, and loads
- Applications—NOS, operations, software, and end-user tasks
- Window/graphic interface—system, library, accelerators, and device drivers
- Disk—controller speed, driver algorithms, cache, and load balance
- Memory—cache type, size, cost, speed
- Database—buffer sizes, lock time-outs, number of users, and caches
- System kernel—base size, efficiency, buffer size, paging, tuning, and configuration
- Executable code—run-time or compile, native or interpreted, environment, file system, and network operating system APIs
- Source code—algorithms, languages, programming method, and compiler

While some people—in particular designers and engineers tuning a large network—find it useful to view the network in terms of its fundamental components, most readers should find it is more useful to view the enterprise network from the big picture. If you see the enterprise network in terms of functional components that seem to create bottlenecks or fight bottleneck fires as they occur, you will miss the structural flaws—the first seven in the list: purpose, wiring infrastructure, redundancy, security, integration, and scalability concerns—that form the basis for Chapter 5 in the enterprise network performance optimization process. Enterprise network performance bottlenecks include these items:

- Design
- Wiring infrastructure
- Security
- Purpose
- Reliability
- Cost
- Speed
- Sophistication
- Environment
- Disk space
- Maintainability
- Time
- Priorities for optimization
- Scalability
- Redundancy
- Integration
- Structural flaws
- Compatibility
- Performance
- Memory
- Functionality
- Network
- Platform independence
- Ease of use/complexity
- Life span
- Organizational culture

Establish performance service levels

Optimizing performance means little without benchmarks and predetermined performance service levels. Service levels are usually a function of response time, processing load (TPS), availability (up-time), and reliability. Other measures are interesting, but usually not actionable. Response time loosely corresponds to *latency*, and processing load corresponds to *bandwidth*. Remember those two irregular terms from Chapter 2? They are important and provide fundamental performance optimization targets for enterprise networks.

Basically, service levels provide a measure for how much work is accomplished and how long it takes for each task to be completed. Most mission-critical operations are measured in terms of reliability and availability. The TPS figure is a ballpark figure, not some quota. Similarly, response time and latency are not absolutes, but represent ranges of performance or an acceptable upper-bound limit. Nonetheless, response is usually calculated as an average value rather than absolute upper-bound. An absolute upper-bound for latency is critical only where real-time control is necessary. Most enterprise network traffic does not control real-time operations. Instead, we look at bandwidth utilization and latency. Figure 3.3 shows enterprise network traffic levels as a function of time. The height of each bar is the bandwidth utilization, while the darkness of each bar represents the performance latency. Note that the color-coding shows where latency exceeds performance thresholds set by the network administrator.

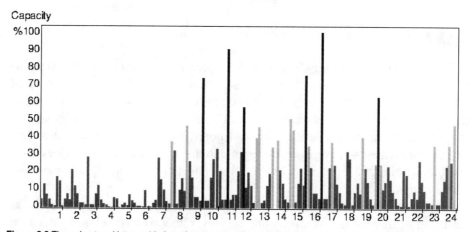

Figure 3.3 Throughput and latency (darker where it exceeds pre-established performance thresholds) over time.

Top hits

The following pie chart in Figure 3.4 indicates the biggest performance hits for the enterprise network.[1] It is strictly categorized so as to produce the fewest

---

[1]IDG, 1993, Framingham, MA.

significant categories. These do not necessarily reflect the root causes of endemic network bottlenecks, but rather reflect what could be repaired quickly. Interestingly, it provides the greatest hope that the performance of the overloaded enterprise can be improved without significant hardware upgrade and bandwidth expansion. This is particularly my experience with enterprise networks based upon workstation and PC technology. Although at least 40 percent of bottlenecks are caused by bandwidth overloads, these causes, and not the ensuing problems, are generally localized to hardware which proved to be defective, mismatched, underpowered, or poorly configured; the causes are not global to the enterprise.

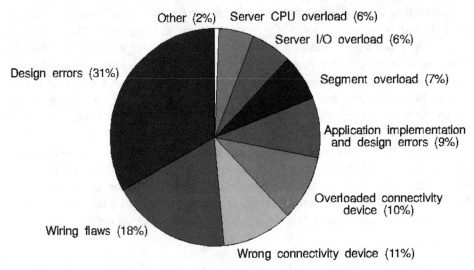

Figure 3.4 IDG's top reasons for enterprise network performance bottlenecks.

Perhaps most surprisingly, many performance problems are directly attributable to flaws in the design of the infrastructure. Typically, the lofty goal of integrating LAN to WAN to host with client/server fail when the integrating applications overload bandwidth and run too slowly. While this might be categorized as bandwidth deficiencies and poor response time, more often than not, the fundamental design of the applications required more services than possible within the limitations of WAN connections or multiple router and gateway hops.

Note that you may want to envision your client/server applications as consisting of more than just clients and servers, rather as channel and gateways; hence "client/channel and gateway/server." Although the transmission channel and gateway may create your bottleneck, review the software design subject to the inherent and *rigid* limitations of this middle component. You may need to test how your SNA or LAN-based applications perform over slower WAN links since these connections exhibit significant latencies and are less tolerant of traffic bursts. This is an issue of *software* optimization and design within the enterprise framework, as discussed in Chapter 5.

Also, while wiring errors (specification violations, mismatched components, and shoddy installation) seem to dominate enterprise network performance flaws, the rate at which they occur has decreased by half since data was collected for *LAN Performance Optimization*. I can certainly dream, but this is most likely attributable to the widespread use of modular network components and standardization of connectors, patch panels, and wiring for data communications. Nonetheless, installers still mislabel lobe cables, fail to check and qualify the run, measure the horizontal cable run distances and overlook the vertical factor, or inaccurately convert between metric specifications and SAE measurements as used in the USA. The newer cable scanners, as described in Chapter 4, are good investments for large networks.

Nonetheless, the top performance hits for the enterprise are derived from latency or bandwidth limitation problems. A handful result from failed integration or configuration. (It was not possible to collate or revise the data within these categories.) However, you might want to look at performance bottlenecks as endemic network-wide problems.

Bandwidth and latency

Chapter 2 presented the paradigms for bandwidth and latency. This section shows you the differences between bandwidth and latency and how they create enterprise network performance bottlenecks. Bandwidth is the available transmission capacity for any network device, channel, or linkage. It is not just really width of the channel, but a factor of the width and sustainable transmission speed. A four lane highway has a greater bandwidth than a one-way street with four lanes in the city because the speed limits differ. Latency is the time required to enter the street or the highway, travel along it, and then exit at the correct destination. Note that the highway does not necessarily constitute a shorter latency than the street because you must get to the highway first; this constitutes a routing issue too. Latency is a factor of all highway travel distance, intersections, and also the travel times on city streets to access and exit from the highway; a complete point-to-point transmission. This is important for data communications and LANs, but particularly so for enterprise networks where there are likely to be multiple hops from intersection to intersection across different channels. Figure 3.5 contrasts bandwidth to latency.

Bandwidth and bandwidth utilization

Bandwidth refers not only to the transmission capacity of the carrier channels, but also the processors, I/O buses, ports, communication processors, and NICs. The protocols in use in structured LAN and MAN environments often include Ethernet, Token-Ring, FDDI, or even ATM in progressive environments. Bandwidth utilization refers to the *engagement of capacity* and is usually represented as a ratio of usage over capacity. However, the differences between Ethernet at 10 Mbits/s or Token-Ring at 16 Mbits/s and FDDI is relatively insignificant. At most, it is ½ an order of magnitude. Even ATM does not represent a significant bandwidth increase at 45 or even 155 Mbits/s. Figure 3.6 shows data bandwidth capacity for common telecommunications and data communications signal carrier methods and protocols.

Connections at 9.6 Kbits/s do not even appear disconnected from the vertical axis. SONET is OC-36. Many high-performance backplanes utilizing routing or switching technology provide anywhere from 1 Gbits/s to 8 Gbits/s in bandwidth. This represents the cutting-edge of transmission technology and dwarfs hub-based backplanes with bandwidths of 240 Mbits/s. Refer to NetSpecs for more information about these and other communications protocols. If your background is from the host environment you may be shocked at the far greater capacities of a LAN protocol; alternatively, LAN experts may not have realized the vast differential between "slow" 4 Mbits/s Token-Ring and the fastest leased line. The enterprise network is rarely as homogeneous as a campus or site network with a hub- or router-based backplanes and FDDI connections servicing a high-rise or several buildings. Host mainframes typically connect to remote terminals at a wire speed of 9.6 Kbits/s. Where PVC lines (DS0) are in-use with capacities to 56 Kbits/s, usually such "speedy" services are attached through DSU/CSUs and multiplexers that share the service among multiple sessions.

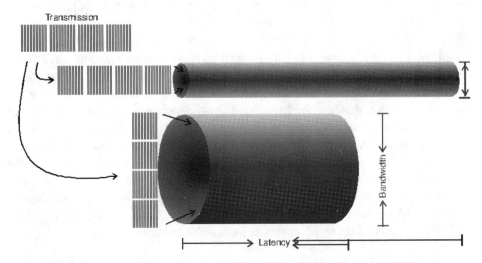

Figure 3.5 The difference between transmission bandwidth (bandwidth utilization) and transmission latency.

There are no clear rules for maximum utilization that is the percentage of your network's bandwidth in use at any time. An Ethernet with more than a few clients degrades exponentially after 35 percent sustained utilization, while Token-Ring or FDDI networks will handle more traffic, but with escalating increases in the latency. The worst-case latency you can tolerate should be expressed as a percentage of the expected server response time and round-trip client/server function execution time. For example, if you expect a quick NFS request to complete in 5 μs, a 10-μs time lag through three routers on a busy enterprise network alone exceeds that. By contrast, that same 10-μs transmission latency is almost irrelevant when compared to an 800-μs delay for a database query.

Measuring network latency requires careful probe placement and some complex math. Publicly available tools like NFSWATCH, NETSTAT, and RCPSPY provide response time figures that include the execution time on the server. And if you keep server statistics, you can extract the server component to calculate network delays. When adding new network facilities, or subnetting existing ones, the network math isn't very straightforward either. For instance, divide a 50 percent utilized Ethernet into four segments without simulation, and you may get Ethernet subnets each with a 40 percent load. So, while decreasing contention for the network lets more packets onto each subnet, you may also create a subnet load greater than the original.

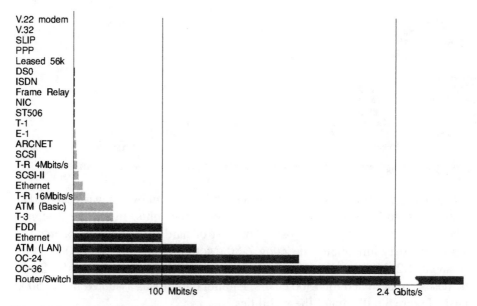

Figure 3.6 Comparison of bandwidth for telecommunications and data communications transmission channels.

A service bundle such as that may provide a 1000 TPS level. This is equivalent to a client/server environment with a 10 Mbits/s or 16 Mbits/s channel and several powerful servers; usually, each server has a capacity of about 40 heavy-duty clients. While it is possible to support hundreds of marginal clients with a server, and typical office-type operations with a superserver and RAID disk subsystems, acceptable performance for transaction processing with client/server peaks out with 20 to 50 clients and 30 to 60 TPS. The technology of CICS and SNA is very different from LAN transactions, so make bandwidth comparisons with some care. Also, do not assume that bandwidth is the limiting factor; it may only appear so. See "Bandwidth myth" on page 75 and "Slow FTP" on page 91.

Note that the DS0 line is barely visible on a scale supporting FDDI. The next graph in Figure 3.7 has an upper limit of 16 Mbits/s to compare analog and SVC digital service in contrast to the typical capacities of NICs, a disk controller, and low-end LAN protocols. Many managers from the mainframe environment misperceive the magnitude difference between a serial link and a LAN link. Correspondingly, network managers with a LAN background may not fully realize the magnitude difference between their Ethernet or Token-Ring throughput and the serial links providing throughput of 1000 TPS on an IBM 9000 mainframe.

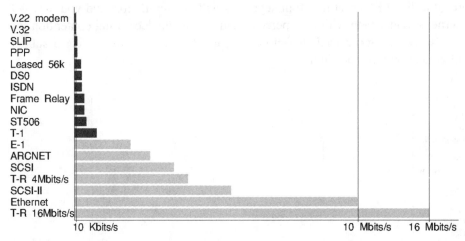

Figure 3.7 Comparison of bandwidth for enterprise networks channels with an upper limit of 16 Mbits/s.

The architectures are radically different. The LAN bandwidth was designed as a connection for PCs with a capacity of about 1000 bit/s. PCs were designed without clear objectives or purposes. The cost of hardware and components has been the primary limitation. The Zilog Z-80, Motorola 6502, or Intel 8086 CPUs ran about 1.2 MHz and could produce about 125,000 calculations/s. Serial channels were designed for 1500 char/s. Parallel output was assumed fast at 5000 chars/s. Figure 3.8 illustrates these abilities.

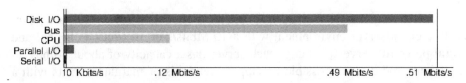

Figure 3.8 A comparison of bandwidth in early PCs.

Although PCs have been compared to mainframes for a decade in terms of performance, you would hardly have considered replacing the $20 million mainframe with a $2000 PC. Although CPU calculation speeds were equivalent, the I/O abilities are different. The mainframe is all bus and I/O. The mainframe was designed for a different purpose, that of processing an organization's data in both batch jobs and real-time on-line processing. The transaction paradigm is based on a small information structure (the record) processed in thousands or millions. The limitation was perceived to be transaction volume rather than the transaction size itself. As such, the host was designed with an enormous addressable memory address space and high internal processing and bus speeds. The front-end processors were designed for these frequent but small transactions and single keystrokes communication; I/O bandwidth was not a design issue. Figure 3.9 shows the bandwidths within a mainframe.

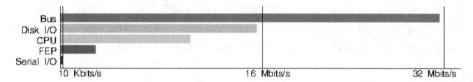

Figure 3.9 A comparison of bandwidth in a mainframe.

Bandwidth utilization is both a factor of highloads, as from CAD or other graphical operations, but it is also from the aggregate of many small loads. Just as a street can fill up with a handful of trailer trucks, it also can fill with a few thousand bicycles. Also, bandwidth is a feature of the number of users too. By increasing the number of users to a segment, you are reducing the shared media bandwidth available to each user, but also (as a baseline) increasing the background traffic level. On rings, each new station decreases access to the token by the electronics delay, the ring monitor, and any basic keep-alive functions. On Ethernet, each new station has no effect except for the basic keep-alive functions.

Mobile and wireless bandwidth

It is important to realize that wireless transmission methods typically use a narrower bandwidth than traditional copper or fiber channels. LAN connections provide about 1.0 to 3.0 Mbits/s rather than the full bandwidth of the Ethernet or Token-Ring on standard media. There is also a significant latency of the transmitting and receiving devices. VSAT usually supplies a channel with 9.6 to 19.2 Kbits/s (with capacity to 128 Kbits/s), while cell-modems are limited not only by the modem technology but also by the quality and reliability of the cellular connection. Loss of a cellular or wireless mobile connection will require connection reestablishment. Some network processes may view the lost connection as a disconnected device. See "A disconnected device" on page 91.

Memory bandwidth

Memory and its available bandwidth is more a system software bottleneck than a true enterprise networking problem. However, when memory on key network devices is overcommitted or not completely freed at the conclusion of a process or task, this will create material distortions in network performance. For example, a C or C++ application may call the malloc( ) library memory allocator and expect the free( ) routine to restore this memory back to the pool of available resources at the conclusion of the application. However, memory allocation does not work that way, much to the frustration of programmers and anyone affected by it. Memory allocation and resource problems are obvious with multitasking and multithreading operations systems where logical pages of (virtual) memory or heaps are mapped to actual physical addresses. Under SVR4 (UNIX), the heap area is called, surprisingly, the *breakpoint* and is manipulated with brk( ) and svrk( ). Paged memory is actually freed by calling the system pager so that unused pages are mapped back into the physical address space. Windows 3.x is not so gracious. A client workstation with corrupted Windows memory heaps (GDI and User) can prevent the release of locked database records on a server with the result that enterprise access is slowed, limited, or frozen.

Latency

Latency is the cumulative delay incurred as packets pass through intermediate nodes, such as repeaters, bridges, routers, gateways, and switches, and the signal propagation time point-to-point between the source and destination. Latency is the round-trip time for a request to be fulfilled and acknowledged over the enterprise network. The latency for a single packet is the amount of delay incurred from when a packet leaves the source, passes through repeaters, bridges, routers, gateways, switches, and the connecting channels until it arrives at the destination. Packet latency *also includes* packet processing time, which is the time to encode and encapsulate data into a packet, and packet transfer time, which is the time required to move the packet to the network itself, and the time a packet may sit in a router queue waiting to be forwarded.

Of note, latency is often measured with single direction (one-way) streams with the result that it is neglible for most intermediate nodes. However, when tested on a bidirectional or backplane environment, latency becomes significant. Vendors rely on incomplete information and do not provide full disclosure on device software and methods so that you can calculate actual latency. While this is true even for NIC cards, bridges, and routers, it is particularly the case with switching hubs and cross-cut switches, where the performance is very dependent upon the software implementation, the hardware method for switching packets, and how broadcasts are resolved on dedicated segments. In fact, vendors cite latency figures for large packets, but rarely for small packets (where latency is more of a problem). It is possible, however, to approximate packet processing rate as the inverse of the packet/s rate for *small packets*, as such:

Packet processing time = 1 / packet/s

You can calculate the packet transfer time with the vendor's single packet latency ratings for large packets. The packet transfer rate is approximately equal to packet size divided by latency for *that* packet size, as shown:

Transfer time = Processing time + (packet size / packet transfer rate)

Usually, packet latency is measured under conditions of low loads. This occurs when the buffers of the bridge, router, gateway, or switch are empty. In a real enterprise network, mixtures of packet sizes, differences in speed of subnets, multiple hops, heterogeneous protocols, and multiple routes cause lengthy queues at the buffers. As such, actual latency through a bridge, router, gateway, or switch usually exceeds the single packet latency in the packet transfer time; often it does so by a considerable margin, depending on the design of the bridge or router as well as the protocols and number of ports supported, the currency of the bridge or router technology and software.

Although latency—transmission channel, intermediate devices, and end nodes—represent a significant portion of the enterprise bottleneck, it is also an important measurement to determine channel growth capacity. In differential calculus you create a derivative to examine the slope of a curve and a second derivative to pinpoint local maximums. In the same way, latency provides a more sensitive handle than bandwidth for determining performance and the critical paths. For example, if latency is static when you increase the bandwidth of a channel assumed to be a bottleneck, the conclusion is that the channel bandwidth is not the bottleneck and nothing has improved. In this case, the transmission time is not a function of the bandwidth. It may be a function of the signal speed, or the intermediate nodes. On the other hand, when latency decreases, you have successfully improved performance.

Measuring latency represents a complex operation with both parallel and dependent paths. Latency for SNA, APPN, or SDLC decomposes to CPU time slices and a consistent delivery from the FEP through a serial link to the display terminal. In fact, excessive host-based latency is usually self-limiting in that processes will time-out or terminate for lack of a synchronized acknowledgment. IPX is also time-sensitive, and excessive latency caused by multiple hops cause connections to time-out.

On the other hand, LAN-based processing is significantly more complex. Transmission latency may not be longer—and in fact may be very much faster. Latency is a factor of server and client CPU time, which represents a fairly consistent service time and a factor of uneven transmission delivery times. Delivery times are often predictable on LANs unless there are significant performance bottlenecks. However, latency on the enterprise network has no self-limiting feature and in fact, overdue transactions usually spawn duplicate requests that merely add to the confusion. In most cases, excessive latency results from peaks, bursts, or the aggregation of normal background traffic. A valid and normal one-way enterprise network delivery time could vary from 0.25 s to 3 s and provide transaction response times as long as 7 seconds. Although this may be unacceptable for an organization, this wide range is normal for asynchronous environments. Figure 3.10 shows the normal distribution for Ethernet latency with a lower latency on a single segment LAN, and a multisegment enterprise network showing the effects of the increased latency.

If this range of 270 µs to 15 s seems excessive, consider that this represents several router hops and maximum collision delays on Ethernet. This results from the cumulative effects of end-to-end latency, as presented in the next few paragraphs. The collision delay component represents most of the latency. This would not be as severe on Token-Ring or FDDI because they are more predictable, however, the routing delay will be disproportionately larger. Figure 3.11 charts latencies for ordinary enterprise network events and a few standard measurements to keep the log scale X-axis in perspective. The range from 1 vs to 3.35 s crosses 10 orders of magnitude and this range is actually more significant than this scale can really convey.

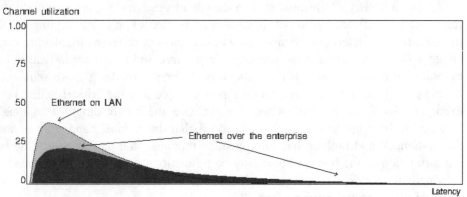

Figure 3.10 Latency comparisons for a single-segment Ethernet LAN and a multiple-segment Ethernet network. Latency for packet delivery ranges from 0.16 µs to 3 s on the LAN, and from 270 µs to more than 15 s on segmented internetworks, enough to cause IPX or SNA time-outs, or increase the traffic levels yet more with duplicate requests.

A less formal treatment of network delay is called the *end-to-end latency*, which is the *cumulative* effect on throughput (for a task such as file transfer) caused by introducing devices between source and destination. End-to-end latency is expressed as a percent of the throughput measured without an intervening bridge, router, gateway, or switch. Unfortunately, end-to-end latencies are influenced by intervening bridge, router, gateway, or switch performance characteristics. Therefore, end-to-end is a meaningful basis of comparison only when measurements are made under highly controlled conditions that match your enterprise environment. In mathematical terms:

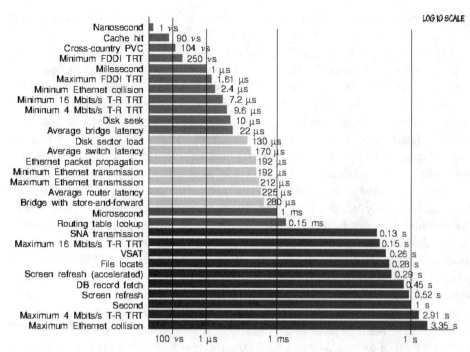

Figure 3.11 Latency values for common enterprise network events and activities (on LOG 10 scale).

$$\text{End-to-end latency} = \text{system latency} / (\text{system latency} + \text{device latency})$$

or

$$\text{End-to-end latency} = 1 / (1 + (\text{device latency} / \text{system latency}))$$

These simple equations are useful because you can see that performance is heavily weighted by the intermediate connectivity device latency. In Chapter 5, where I discuss selection of bridges versus routers, routers vs. switches, other devices or architectures, you will want to contrast overall packet processing time with overall packet transfer time and the component of that transmission time lost in the connectivity devices. You will also want to analyze packet sizes and overall loading as this alters the utility of bridges, routers, or switches.

In spite of its limitations as a performance benchmark, end-to-end latency is a useful measurement. It closely parallels user response time for such activities as screen updates and file transfers. For example, a 50 percent end-to-end latency would mean that a user would have to wait twice as long for a file transfer or a screen update to occur if only a single connectivity device is introduced in the path. If a path involves several hops through bridges, the effects of end-to-end latency compound very quickly. With 90 percent end-to-end latency, four hops leave a reasonable 70 percent of bandwidth. However, at a lower 60 percent end-to-end latency, the cumulative throughput drops to 26% of available bandwidth. True wire speed would be 100 percent of bandwidth and 100 percent end-to-end latency under ideal conditions. You are going to see the effects of end-to-end latency bottlenecks repeated implicitly throughout this book.

Figure 3.12 revisits Figure 3.5, but with a difference. This illustration compares two types of latencies. Figure 3.12 compares latency for ATM providing bandwidth of 622 Mbits/s on SONET with a digital connection rated DS0 providing only 56 Kbits/s in bandwidth. Although these transmission mediums are different, the point is that signal propagation speed is constant for most media. Signals propagate at a major fraction of the speed of light—in fact the SONET carrier may provide transport for both the cross-country ATM signal and the lowly analog modem line—and nonetheless provide identical transmission latencies (once the differences between call setup and breakdown times are factored out). Any intermediate latencies in the transmission cloud provided by the command carrier facility are identical; you might note that the transmission cloud will impose random SONET and SMDS switching delays for all traffic unless a virtual point-to-point circuit is established. Albeit, the ATM provides 10 times the traffic throughput, but I want to make perfectly clear that the transmission latencies on the wire are identical. You might note that the modem signal from point-to-point might actually be faster because the intermediate and terminating switching electronics of each ATM switch has latencies of about 170 μs (for a total of *at least* 340 μs), while the modem adds 100 μs at each endpoint.

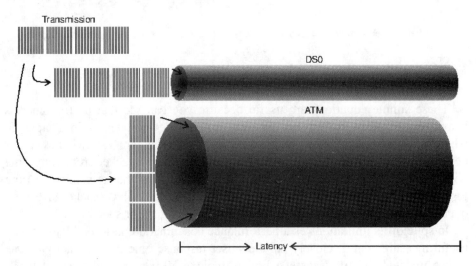

Figure 3.12 Latency comparisons for ATM (on SONET) and DS0 (on SONET or copper).

Latency is added into the transmission time by any intermediate device. It is greater for devices that buffer incoming transmissions and hold them while earlier (or prioritized) transmissions are forwarded. This latency is typical with any device that has a store-and-forward buffer or feature. This is typically seen with SNA, high-bandwidth backplanes, routers, and switches. The flow control mechanism (or *pacing* as IBM calls this) when uneven, creates long response times and session failures. Print jobs and transfers with SNA protocols tunneled into TCP/IP can hog so much bandwidth that no interactive traffic can get through. Even adaptive pacing, which negotiates the window size and transmission burst lengths, is not always compatible with the enterprise network.

As a last indignity, consider the delays for each additional hop and recall how each transmission in Figure 3.10 dramatically extended the normal latency range for multiple hops. Do not forget to consider that most enterprise network transmissions are not unidirectional; the acknowledgment message, server processing time, and fulfillment are usually part of a sequential (synchronous) operation on the client and elapsed response time will be the sum of all multimessage latencies. The entire sequence is especially relevant when assessing bottlenecks in client/server applications propagated to remote sites or through multihop connections.

Mobile and wireless latency

Wireless and mobile computing use different transmission channels. It is important to realize that while wireless transmission methods typically use a narrower bandwidth than traditional copper or fiber channels they also increase the latency of the transmission because each transmitter and receiver is also effectively a receiver and router. Expect latencies in the neighborhood of 300 µs. Mobile computing typically uses VSAT, radio, or cellular radio technology. Latencies may be as long as several seconds, but unless connections are always maintained in an active mode, each call will also require a setup time from 2 to 15 seconds. Loss of connection requires call reestablishment. Some network processes view lost connections as a disconnected device. See "A disconnected device" on page 91.

Network load

The fundamental cause for enterprise network bottlenecks is the traffic load imposed. Network load is not always *actively* imposed, particularly in case of a large enterprise network where a considerable amount of traffic is related to *passive* network housekeeping, management, security, backup, and processing messaging. The following list indicates some of the most likely sources for traffic:

- Transaction processing
- Network user access authentication
- Network login
- Remote file or program loading
- Node data backup
- E-mail
- Network management
- Server and host backup
- Routed traffic

Without traffic there is no load. Without a load, there should be no bottleneck. Just as metropolitan highways and transit systems are virtually empty at midnight in contrast to the peak commuter load, the enterprise network channel utilization should be zero (or very low) without an external traffic load. Ideally, channel utilization will be zero. Otherwise, background loading indicates runaway processes, poorly configured connectivity devices, SNMP and other in-band performance monitoring or management tasks, tasks initiated by time clocks, hackers, illegal traffic, parasitic traffic, or errors. Chapter 4 discusses the tools useful for detailing the sources of background traffic noise. Because content is so important, you really need to know what is insignificant, extraneous, or critical to the underlying purpose of enterprise network. Figure 3.13 plots LANDesk information as daily traffic loads as a function of time and content.

Although Chapter 5 clearly makes the point that the best approach for relieving enterprise network bottlenecks and overloads is to lighten the traffic load, most performance optimization occurs in anticipation of (or as a negative reaction to) some fundamental design and infrastructure changes. Typically, additional users, new applications, an upgrade from DOS to Windows, a shift from text-based to graphical user interfaces, or full-scale deployment of data-, graphic-, or image-intensive processing applications create additional traffic burdens. Although the increased traffic may raise bandwidth utilization on channel, connectivity, and processing devices, the increased load usually results in increased latency and decreased user response time. You may see saturation on specific devices or segments, delays for processing, and even network panics.

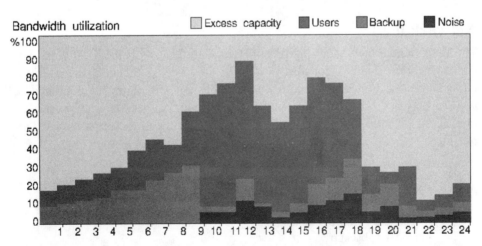

Figure 3.13 The traffic as a function of content.

Note that as traffic bandwidth approaches capacity, the slope of latency curve goes horizontal; as traffic exceeds capacity, wait for transmission will be infinite. It usually disproportionately increases latency, as Figure 3.14 shows.

Bandwidth utilization

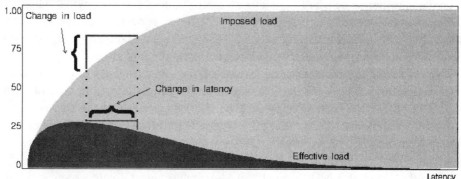

Figure 3.14 The traffic levels as a function of content.

While it is ideal to run benchmarks or model increased traffic loading (as described in Chapter 4) in anticipation of such changes, many planners do not anticipate the affects of increased bandwidth utilization or latencies associated with mixed protocols and conversions, multiple hops, and routing. While synchronized traffic (such as SNA or SLDC) is well-behaved, traffic from LAN-based protocols is not. Traffic growth over time is normal. New users, additional subnets, new remote sites, host and LAN integration, new applications, application functionality creepage—who ever heard of a new release or upgrade that does less and runs faster?—increased computerization, and more complex applications contribute to increase CPU and enterprise loading. Figure 3.15 represents the expected bandwidth growth in a typical enterprise environment.

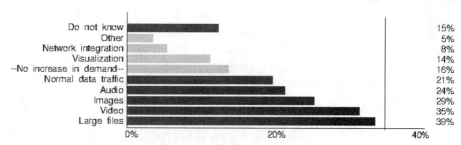

Figure 3.15 The types of traffic demanding more bandwidth.[2]

Applications themselves can create bottlenecks. Many are simply not designed for the long latencies and congestion of the multiple-hop enterprise network. They are not optimized for the narrow bandwidth, lack of tolerance for LAN synchronization, and traffic bursts over remote WAN connections. Others suffer from age deficiencies, pure and simple. For example, time and date stamping and the formats applied are expected to create a $20 billion conversion

---

[2]Business Research Group, Newton, MA.

effort near the turn of the century. Diverse date and time-stamp formats represent a conversion bottleneck as LANs are connected into sites worldwide. This is a simple format bottleneck, because dates and times generally require the same space regardless of format. However, characters and data are usually more restrictive, and represent a greater risk to enterprise organizations.

Network load is more than a feature of the operating system or network operating system. The workload charactization is a function of the type of work too. For example, NFS workloads are typically at opposite ends of the spectrum. You either have small files (minimum of one 8 kb block) and an abundance of file lookups and getattr workloads (they correspond to 90 percent of the system load) and thus attribute-intensive, or you have the image processing with file sizes in the range of 3 MB to gigabytes. Attribute request create server and NIC bottlenecks, whereas large files will obviously tax the infrastructure bandwidth.

With the transition of LANs and host environments into a true multinational enterprise network, you may need to consider the ramifications of language and the need to adopt Unicode (ISO 8859-1) character set for internationalization and integration. However, converting 7-bit EBCDIC or 8-bit ASCII characters into 16-bit or 32-bit representations comprises a logistical, performance, storage, and transmission bottleneck.

Shared media

Bandwidth on all LANs is shared by the protocol access method. Ethernet applies a first-come, first-served basis modified by collisions that cause devices to retry after a random exponential multiple of 9.6 µs. Token-Ring, FDDI, and ARCNET split the bandwidth by passing a token around the physical or logical ring. Typically no station gets more than the THT (that represents the maximum frame or packet size) every TRT. In other words, each device shares about THT / TRT of the available bandwidth on the media. No device gets the full bandwidth unless it is the only device on the network—and that does not make much sense. Devices rarely get more than the bandwidth for more than a packet or two; the rest is shared among other devices. Some people divide the node count into the bandwidth to determine an *approximate* shared resources available for each node that does not include the burstiness of network traffic.

As a result, LANs (and enterprise networks) experience bottlenecks when the aggregate demand for channel bandwidth exceeds the available bandwidth. Lack of adequate shared bandwidth creates both increased latency and bandwidth saturation. Ethernet saturation not only means that bandwidth is fully utilized, but saturation also increases the transmission latency, and (somewhat counterintuitively) also decreases traffic throughput. Ring protocol saturation is not as disruptive; as channel utility (and traffic throughput) approaches 100 percent, the latency increases to the theoretical protocol maximum TRT. You certainly know this; everyone who works with LANs believes that LAN performance suffers because the wire is overloaded.

Bandwidth myth

Insufficient bandwidth is an overrated bottleneck; it is not the most common bottleneck for the enterprise network, and certainly not even for LANs[3]. Bandwidth on the transmission channel is a bottleneck for mixed digital communication and WAN connections. On LANs and enterprise networks, bottlenecks at the intermediate nodes are usually the active constraints.

Differentiate these bandwidth bottlenecks from channel limitations. Before you assume that bandwidth through a CPU bus is a limitation, make certain that the bus and the adapters are configured for the full bandwidth utilization. It is possible with PCI, VESA VL, EISA, ISA buses to mismatch the CMOS or BIOS configurations with the result that throughput is 10 percent or less of bus capacity. It looks like a bottleneck, but it is only a functional one. Although attempts to upgrade may actually resolve the problem, you have skirted around it and not really addressed it. For example, I have experienced integration problems with several SCSI and IDE cards yielded disk and NIC throughput under 1 Mbits/s on different Pentium 60 MHz systems. While I might have replaced these systems with 90 MHz or 100 MHz machines, a solution was within the original hardware configuration. Similarly, a LaserMaster WinJet card in a 90 MHz Pentium provided network print output at speeds slower than the original Intel 386 20 MHz system due to I/O configuration problems. In all cases, Pentium I/O throughput was improved by excluding memory ranges, setting new IRQ addresses, and creating new base memory addresses. This solution was neither simple or quick, but was neither expensive nor required upper management signatures for more hardware. On the other hand, CPU, I/O, and NIC performance and new technology applications *are stressing* the shared media bandwidth, particularly for videoconferencing, medical imaging, multimedia, graphics (CAD/CAM), and remote-site hierarchical storage management (HSM) or recovery operations. As such, you will want to measure, test, and model bandwidth capacity and utilization. Chapter 4 demonstrates that this is easily achieved.

Interconnectivity devices

Networks consist of links and nodes. Nodes can be *end* nodes, like clients and servers, or *intermediate* nodes, like hubs, bridges, repeaters, routers, gateways, or switches, which are the interconnectivity devices. The LAN began as a single network with shared media. Now, the enterprise is a conglomeration of LANs with end nodes, interconnection devices as intermediate nodes, and many other global resources and processes. Design is the significant enterprise network performance topic. Refer to Figure 3.3 for the top ten hits. Although selection of the wiring infrastructure and protocols obviously affect available bandwidth and data communications latency, transmission latency is the same magnitude for all media,

---

[3]*Debunking the bandwidth myths*, Ed Krol, *Network World*, p. 16, April 25, 1994.

wiring (and fiber) variants, and protocols. Clearly, increasing traffic load with new applications increases the base-level bandwidth. This in turn adversely affects latency, all the more so when the traffic load already uses most of the available transmission bandwidth. Even simple queueing models show this truism.

The latency effects are actually more profound for the connectivity devices that connect these transmission channels because they are more complex, because they are active, and because each hop requires a signal acquisition slot for the continued transmission. The devices that connect segments and hosts into the enterprise include:

- NICs
- Repeater
- Bridge
- Switch

- MAU or wiring concentrator
- Hub
- Router
- Gateway

The performance characters vary for each type of device and also by vendor. Because most connectivity devices have a software—or at least a firmware—component, vendor implementation, integration, and conformance to protocol specifications add an unpredictable performance component. Performance varies as a function of configuration, loading, and installation.

For example, consider router vendor claims from manufacturers cisco and Wellfleet of forwarding rates from 70,000 packets/s (pps) to 500,000 pps, as corroborated by Interlab. I picked these vendors because together they represent about 80 percent of the enterprise router market. In order to achieve numbers like these, the test methodology and environment is represented by a unidirectional traffic load, empty buffers, and a traffic load generator such as PERFORM3 (described in Chapter 4). This workload characterization is very different from most enterprise network environments where mixed traffic loads, multiple protocols and a distribution of packet sizes (mean 170 bytes with an asymmetrical distribution to the smaller packet sizes) bombard the many hubs, bridges, routers, switches, and gateways. Typically, these devices confront a bidirectional or even multidirectional traffic load, and also provide filtering based upon SNMP community strings and OSI gateway functionality by encapsulating or translating protocols. Performance is affected by configuration issues, such as the total number of ports and number of active posts. As such, router total throughput can drop to 3000 pps in a typical enterprise, yielding widespread bottlenecks and an enraged user community.

Because it is not realistic to compare router performance from different vendors over all possible configurations, traffic loads, and protocol mixes the next few sections describing the major types of connectivity devices, their performance characteristics, and potential bottlenecks.

NICs

The network interface card can create significant bottlenecks on the enterprise network, usually at servers and home-brewed gateways. The implementation of NIC and network drivers create CPU load, in most cases a minimal load, but in the adverse situation, the overhead is significant. This is particularly prevalent when the node has multiple NICs or bus mastering units. When this load overtaxes the CPU, the result is to slow disk I/O and fundamental processing and the network node becomes the bottleneck, not the bandwidth itself. Chapter 5 discussed solutions for NIC-related bottlenecks.

MAU or wiring concentrator

The MAU, MSAU, or wiring concentrator is more specifically a part of LAN (10Base-T, 100Base-T, 100BaseVG-AnyLAN, Token-Ring, or ARCNET) infrastructure than it is part of the enterprise, because this really is a passive wiring connection for constructing a LAN segment or subnet. It provides a solution to the series wiring of 10Base5 Ethernet in that a wiring closet-based solution (such as Token-Ring or 10Base-T) is generally self-healing. In other words, when a node, lobe cable, connector, or connection fails, that port fails alone without disabling the entire network segment or subnet. However, this passive device may be included as part of the enterprise as the primary focal point for attaching a LAN segment or subnet. Wiring concentrators do not have logic circuits more complex than those required to disconnect failed or cascaded units.

In fact, the wiring concentrator becomes a bottleneck when aggregate load for attached devices exceeds the availability bandwidth of the shared media. Furthermore, when too many devices are attached to the wiring concentrator, or when they are daisy-chained beyond node count limitations, TRT-based latency and collision-based latency will create LAN bottlenecks. Depending upon the integration of such a LAN into the larger enterprise, these problems can create traffic gridlock or a network panic. If a designer were to somehow create an enterprise based around a wiring concentrator—it is logically and by definition inconceivable—the shared media bandwidth limitations would render the network dysfunctional.

When viewing Ethernet in contrast to Token-Ring (or FDDI), it is also important to realize that Token-Ring (and FDDI) suffers more from jitter than self-synchronizing protocols like Ethernet, as both *LAN Performance Optimization* and *Computer Performance Optimization* addressed. These relate to near-end crosstalk and impedance mismatches in cabling (and jumpers). Jitter is proportionally related to the length of the ring, the number of nodes, and the number of bridges or routers. Because jitter is cumulative, it has a particularly profound effect on the enterprise network. When jitter gets really bad, nodes and servers crash and precipitate network panic. You may see vendors providing solutions for jitter in the form of *active, nonretimed electronics, tank* or *ringer circuit retiming*, and *phased-locked loop reclocking*. Hub- and switch-centric designs help resolve these limitations of ring configurations and token-passing protocols by retiming transmissions at every repetition.

Repeater

A repeater is a device used for replicating a decaying transmission signal so that it can be sent to a node or network device that is beyond normal reach. Repeaters were the first intermediate LAN nodes. Repeaters operate bit by bit at the physical level, and are protocol independent. Packets sent on any one LAN cable segment are repeated indiscriminately on *all* others, creating unnecessary traffic. The repeater has no positive performance effect except to make an excessive design conform to protocol specifications. It can also boost poor-quality signals and enhance the chance for bad CRC, corrupt data, or other errors.

However, the repeater is a vestige of Token-Ring or 10Base5 LANs that were expanded for site, campus, or quasi-enterprise connectivity before the idea that aggregate traffic loads could ever exceed shared media bandwidth was even conceivable. The repeater also provided the means to actually create Ethernets with 1024 devices, the legal maximum. This is antithetical to microsegmentation, a key method for optimizing LAN and enterprise performance. Now, the repeater is a bottleneck just waiting to happen, or an active bottleneck that you haven't discovered yet. It aggregates the load for attached segments and increases the overall bandwidth utilization. The repeater adds to the TRT for ring-based protocols due to its internal and lobe electronics delay, increases the window during which an Ethernet collision can occur, and must acquire the channel (via token or lack of collision) for every transmission. If it cannot acquire the channel for outgoing signal repetition, it drops the packet. It drops packets that arrive simultaneously, and also during the processing overlap interval. It is also a bidirectional device. When you model this device, it has two source distributions, two output distributions, but only a single processing mechanism. The latency on a repeater is about 0.02 to 100 µs. Note that every NIC on a Token-Ring or FDDI network functions as a repeater.

Hub

The hub is an active wiring concentrator that repeats transmission signals between end nodes. It was a natural extension of the repeater as an intermediate node and infrastructure builder. Because hubs (repeaters) do not look at packet addresses, repeating hubs repeat incoming bits on all ports. This is great for multicasts and broadcasts, but inefficient of linked bandwidth for the majority of packets addressed to a single end node. The hub fulfills the same function as the MAU, MSAU, or passive wiring concentrator and even creates the same bottlenecks as wiring concentrators and repeaters.

However, the hub provides several degrees of freedom in LAN design by allowing longer distances between node and hub or the use of inferior lobe wire. Some managed hubs provide an in-band or even out-of-band (true side-band with serial connection) capture of SNMP statistics or configurable traffic and error rates. Some hubs have backplanes that provide increased bandwidth. These devices really are an inventive stuffing of many hubs into a single package that provides manual or remotely configurable microsegmentation. There is nothing inherently different in a hub with multiple channels from a performance standpoint other than it is easier to reconfigure a LAN or multiple LANs and balance bandwidth between two or more shared media segments. Hub performance varies with configuration, and how effectively you can microsegment the network. Refer to "Interconnectivity devices" on page 75 to review the bottlenecks that can be caused by hubs in an environment characterized by chaotic enterprise network workloads.

On the other hand, some hubs include routed or switched backplanes. These are not hubs per se, but rather hybrid devices installed for simple configuration into a wiring concentrator. You will need to analyze the performance ramifications on a per device basis.

Bridge

A bridge is a signal repeater that filters traffic at the MAC level. It was next in the succession of intermediate LAN nodes. Bridges momentarily buffer entire packets, and improve upon the repeater (which would drop packets arriving at the same time or during the processing overlap interval). Like the repeater, it is also a bidirectional or even multiport device. When you model this device, see that it has (at least) two source distributions, (at least) two output distributions, a single buffering mechanism, and usually only a single processing mechanism. Collecting source addresses, bridges build tables that indicate which nodes are on which of the bridges' ports. The bridges use these tables to look up which of their ports a packet's destination node is on. This means that traffic sent between nodes on a shared LAN segment will be filtered (at the MAC-level) and not be wastefully forwarded to other subnets or segments. In networks with many nodes, bridges must store and search many addresses, making them expensive, slow, and wasteful of linked common bandwidth.

Although the bridge is a repeater, it is slower because it buffers packets. This does not mean that the bridge is not useful. In fact, because it buffers packets rather than simply dropping the overrun, it can decrease overall network traffic throughput and decrease latency. While it may degrade the results for some specific applications, the overall effect is to better internet performance. The bridge typically reduces the load passed between enterprise segments by creating a firewall based on MAC-level source and destination addresses. However, it can create extra and unnecessary traffic; improper configuration of a bridge causes it to function as a repeater and aggregate the traffic levels to all attached segments.

Also, a bridge requires just as much overhead and CPU processing power to forward a small packet as it does to process a large packet. As a result, because the distribution for packet sizes of internetwork traffic is 95 percent or greater between 24 and 127 bytes, and almost 80 percent of all packets are 32 bytes or less in an enterprise network environment with SNA traffic, the bridge represents a bottleneck and source of latency on the order of 100 to 300 μs.

Because there is significant logic required to filter and route traffic, the bridge adds latency from both the internal electronics delays, buffer delay, and signal acquisition time. If those packet buffers overflow, packets may be dropped. This has the unfortunate tendency to increase enterprise traffic as clients reissue the unfilled requests and servers retransmit unacknowledged services. Lastly, most bridges are not very sophisticated or advanced in hardware or software. The bridge rarely has room for more than 4000 addresses in its routing tables. If the enterprise network supports more addresses than that, expect to see a high load of unnecessarily repeated traffic, lost packets, and dropped packets. Review "Inter-connectivity devices" on page 75 for the connectivity bottlenecks that can be caused by bridges in an environment overstressed by high enterprise loads.

Router

The router has been the core interconnectivity device for building enterprise networks for the last six years, and is also an intermediate node. Routers store and forward like bridges, but they also process protocols at the data and network layers of the OSI model. However, unlike bridges (which build and search address tables), routers use compact information supplied by protocols to forward packets to their destinations. There are literally thousands in place to provide attachment and interconnectivity for the Internet. The router provides traffic filtering, routing, and forwarding; transport- and network-level transaction; and security and traffic firewalls to prevent a network panic from overloading individual segments. However, it is not free from the effects of bottlenecks, or too perfect to spawn its own performance bottlenecks.

Also, a router requires just as much overhead and CPU processing power to forward a small packet as it does to process a large packet. As a result, because the distribution for packet sizes of internetwork traffic is 95 percent or greater between 24 and 127 bytes, and almost 80 percent of all packets are 32 bytes or less in an enterprise network environment with SNA traffic, the router is typically a substantial bottleneck and source of latency. It is very susceptible to packet size, input source and output loading, and the complexity of the network-level protocol translations. Like both the repeater and the bridge, the router is also a bidirectional or even multiport or multiprotocol device. When you model the router, realize that it has (at least) two source distributions, (at least) two output distributions, some type of buffering mechanism that may be singular or a rotary as described in Chapter 2, and may have multiple processing mechanisms. As such, the router may represent a multiple server, multiple queue, or a multiple output queue service mechanism, and is more difficult to properly simulate in a queueing or steady-state model, as described in Chapter 4.

The router improves on both the repeater and the bridge because it filters traffic at the MAC, data, and network levels. This is not without cost in terms of transmission latency. The effects from dropped packets and overruns depends on input packet arrival times and packet sizes, as well as the output segment loads. It is useful as a firewall to segment traffic. Some vendors have implemented protocol spoofing and protocol translation or encapsulation for hetergeneous networks. *Spoofing* is the process of matching the protocol messaging requirements (particularly for IPX and SNA) so that sessions do not time-out. *Translation* is the process of converting one protocol into another for transport, while *encapsulation* is the process of embedding one protocol inside another as payload for transport. These three processes are software-intensive. They should perform well until the traffic arrival exceeds processing time, but I suspect that the processing time is fairly lengthy, with the result that it will be some time before router CPU power measures up to the enterprise network.

Because the router employs a more complex mechanism for analyzing packets—and it must decode the packet to extract the LLC, IP, or MAC destination address from the data field—it is much slower than a repeater, and far slower than a bridge, with latency from 200 to 500 μs. Routing protocols vary enormously. There are at least twenty different important LAN routing protocols and each has a tradeoff in terms of bandwidth utilization, transaction time, simplicity, reliability, latency, integration, and compatibility. Routing forwarding speeds vary enormously from the vendor claims to actuality in your environment; refer to "Interconnectivity devices" on page 75 to review the bottlenecks that can be caused by routers stressed by high enterprise traffic loads.

Do not underestimate the gridlock potential from a router with buggy software. This is a common problem, and more common is the issue of integration and software upgrades and the adverse effects on router software. While bugs in the firmware or software will usually not create a packet storm or cause a network panic, you are likely to experience performance degradation on order of 5 to 20 percent of enterprise bandwidth. These flaws are neither constant nor endless; often you can interactively debug problems to sidestep performance hits.

The most important routing bottleneck for enterprise networks is the router's support for different implementation of high-speed LAN protocols. As many network managers apply Fast Ethernet and FDDI band-aids to overloaded LAN segments, the interconnectivity and interoperability into the enterprise is really an unknown. Will your routers handle 100 Mbits/s Token-Ring, 100Base-X, 100Base-T, 100BaseVG-AnyLAN, other variants, or new protocols? The most *obvious* important routing bottleneck is the broadcast storm routers create when they check the validity of or rebuild these routing tables. When the routing tables are corrupted and abandoned, the traffic may be simply repeated; this can create a network panic as the bandwidth everywhere on the enterprise is flooded with misdirected and progressively duplicated packets.

Multiport router

The multiport is typically seen as part of a site or campus wiring hub. It is sometimes called an *enterprise hub* or an *enterprise router* because the vendor is seeking to sell an integrated package for internetworking in vertical buildings or large organizations. It is incorporated with a high-performance backplane and provides dynamic configuration of microsegmented LANs. When you model this device it really is nothing more than a multiprotocol, multiport router. It may have a high-speed backplane channel that connects microsegments with bandwidths from 150 Mbits/s to 8 Gbits/s. You need to partition this device into separate routers with a special transmission link. Alternatively, you can simplify the model by assuming no bandwidth or latency limitations between individual router processors. The key issue is to see this device as a collection of attached intermediate devices, rather than as a single unit. Many vendors do this with the result that the multiport router appears to provide insignificant latency and unlimited bandwidth. This is certainly not the case. While some multiport routers are faster than router software on a PC, the performance differential is minimal.

While these high-end devices usually have sufficient bandwidth to handle at wire speed as many segments as can be physically attached, the management and configuration of these very complex routers remains one of the foremost bottlenecks. It is not so much as a performance bottleneck, but rather a technical one. When an enterprise network is built around multiple sites, each supported by a central multiport router in a hub, the most significant bottleneck is remote management and configuration of the router to minimize the cross-traffic.

The primary multiport router performance bottleneck is caused by single port overflows or bottlenecks. Because it is a singular focal point, the failure of a primary channel, port, power supply, logic circuit, or backplane will gridlock that complex traffic intersection. Some multiport routers have redundant power supplies, duplicate backplanes, spanning tree capability, and SNMP support. However, one problem causes problems on the other ports, backs up traffic, fills the incoming packet buffers, which creates bandwidth bottlenecks, dropped packets, and enduring latency lags. Also, the spanning trees are not instantaneous; the hot standby unit or route does not kick in instantly. It could take many minutes, at which point SNA sessions and IPX connections will time-out.

By the way, a spanning tree configured with multiple routers also can create a real bottleneck on the enterprise if both routers are simultaneously active. The switchover when one path fails is neither instantaneous, overhead-free, nor failsafe. The time required for a router to recognize the route failure and reconfigure itself is called the *router convergence time*. It is a factor of the router protocol(s), the traffic levels, and the size and complexity of the enterprise. A lost route must be noticed for what it is—router convergence usually requires several minutes— and requires that the standby router construct and optimize its own routing tables based on its active routing protocol.

For all bridging and routing designs, a fundamental flaw is to badly group servers, hosts, and their clients on different network segments, as shown in "divide a 50 percent utilized Ethernet" on page 63. This has the result of adding unnecessary traffic to highway backbone and increasing latencies. When servers and clients cannot be localized, the performance hit may render the application unusable. This need not be. There are other methods of supporting distant and/or remote clients without a significant hit, as Chapter 5 details.

Gateway

An OSI gateway typically provides conversion for interconnectivity and interoperability between a host and LAN environment at all levels of the OSI network model. The gateway is some combination of hardware and software that provides access for workstations and mainframes to share information. The gateway was also the first enterprise network node, but in a different configuration, that of a dual-ported node talking to two or more subnets or segments. This configuration provided the means to microsegment overloaded LANs, to split transmission channel bandwidth, and to distance the effects of busy file servers from each other. Typically, for this configuration the file servers also were the gateways, which required additional NICs, a considerable amount of specialized configuration and thoughtful load balancing, and the same design limitations that bridging and routing technology cause now. It created all the bottlenecks of repeaters, bridges, and routers when improperly configured. Additionally, it added overhead and latency as it converted or encapsulated one protocol inside another.

The gateway is a bidirectional or even multiport device. Its reason for existing is usually multiprotocol translation or encapsulation. When you model it, realize that it has (at least) two source distributions, (at least) two output distributions, some type of software and hardware buffering mechanism which may be singular or a rotary as described in Chapter 2, and may have multiple processing mechanism. Because it may be multiprocessing, consider it a multiple server, multiple queue, and multiple output queue service mechanism. It is more difficult to simulate in queueing or steady-state models, as described in Chapter 4.

This initial use for the gateway has been pushed down to the high-performance routers. More often, now, you will employ a gateway to provide connectivity from a host to PCs on a preexisting LAN. This device—and also just software—represents the potential for integration and compatibility bottlenecks. Additionally, gateways should provide bidirectional functionality. This is not always the case. The gateway may provide one-way access from the PC to the mainframe, but not reliably the other way. Also, the PCs on the enterprise may be able to use host-based printers, but the mainframe may not be able to converse with the graphics printers through the gateway on the enterprise. Refer to "Interconnectivity devices" on page 75 to review the bottlenecks caused by gateways in an environment characterized by high enterprise network loads.

Bandwidth of the hardware and the software also represents a significant issue, as a gateway may provide good service for one or two trial users, but none when rolled out to the enterprise. The gateway may not provide the complete network and process integration as envisioned. A typical application for a gateway is E-mail service and routing. While it may be possible to actually grab mail from one system to the other through a gateway, integrating the mail messages into the database of the other may not. It also may not be possible to know when to purge messages from the source as these mail gateways are not always fully reliable.

Switch

The switch is the newest kind of intermediate LAN node, a router, if you will, built in hardware rather than software. The switch is represented by at least two different products. There is the *matrix switch* or *network switch,* which provides automatic configuration of segments and subnets; there is also the *virtual switch,* which constructs temporary connections much like a PBX. It is important to realize that these are two very different implementations of intermediate connectivity devices. The matrix switch is a microsegmentation device—a wiring concentrator, as shown in Figure 3.16.

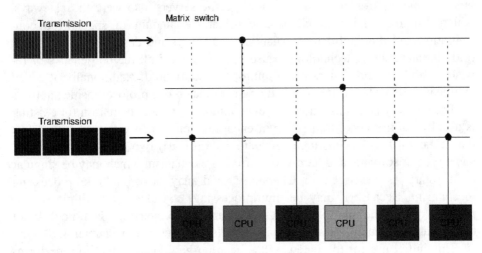

Figure 3.16 A matrix switch provides microsegmentation.

It is very useful in hubs and backplanes, whereas the virtual switch dedicates full media and protocol bandwidth to connections. The bottlenecks created by matrix switches are the same as are inherent in any scheme for microsegmentation. Primarily, it is a shared media technology and eventually there is not enough bandwidth to support each LAN and provide interLAN communication.

On the other hand, it is important to differentiate the matrix or network switch from the newer technologies. The virtual switch takes microsegmentation to the ultimate by including only the source and destination nodes in a temporary network, as shown in Figure 3.17. This provides full protocol bandwidth to the connection. The connection lasts only as long as necessary to transfer the cell, frame, or packet. Virtual switching creates bottlenecks because the switching and connection is not instantaneous. Switch latency for LANNET MultiNet is 13 µs, Atlantec PowerHub is 120 µs, while Triticom SwitchIt was 475 µs. All concurrent attempts to communicate with the connected nodes results in a busy signal, dropped packets, and added network loads created by the recovery or retransmission attempts. This is called *blocking*. When blocking is buffered so as to alleviate it, the buffer delay adds to the switching latency.

Vendors too often do not understand the performance ramifications of their switch designs and also are caught in the media pressure hyping switching technology and its benefits for improved bandwidth and network performance. They do not want to detail the exact switching modes and algorithms because performance is very similar to routers—in many cases. Refer to "Interconnectivity devices" on page 75 to review the bottlenecks that can be caused by switches in an environment characterized by extreme multidirectional enterprise network traffic loads. In fact, the following two illustrations should convey the inherent opportunities and performance limitations in a switch-based design.

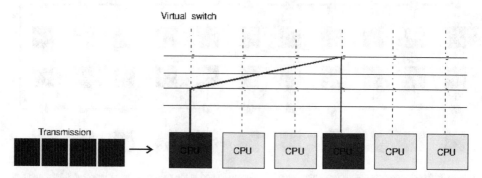

Figure 3.17 A virtual switch provides virtualized microsegmentation for paired transmissions and full bandwidth to that connection.

Figure 3.18 shows a traditional Ethernet segment.

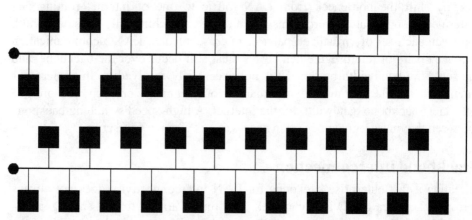

Figure 3.18 This logical illustration of a typical Ethernet single-segment LAN represents a frequent bottleneck with adverse performance feedback to the larger enterprise network. It does not matter whether the wiring infrastructure is 10Base-2, 10Base-5, or 10Base-T; the performance effects are the same although the 10Base-T provides a simpler infrastructure upgrade path.

Whereas, Figure 3.19 shows the substitution of a switching hub or switch infrastucture. The pairs of nodes could also represent microsegmented LANs. I have kept the logical wiring diagram the same so that you can see that the switch merely connects pairs of nodes that must still communicate over some sort of backbone. Any fundamental bandwidth improvement (as with duplex Ethernet) or traffic containment is limited to the paired nodes or microsegmented LANs. Most PCs lack the ability to transmit more than 1 Mbits/s—excepting high-end 486 or Pentium workstations—and will perversely experience longer latency and new backbone routing delays.

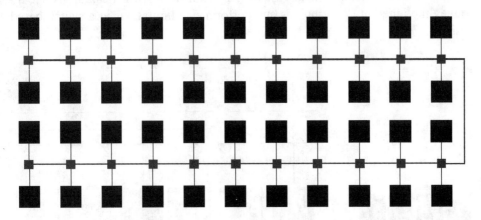

Figure 3.19 Switched network microsegmentation benefits are limited to the paired nodes or localized LANs, and any traffic between these segments bound to other segments faces the speed and bandwidth limit of the backbone.

My point is that switch technology (or even duplex switching and network virtualization) is not a solution just by itself. In fact, switching technology may merely shift the bottleneck from LAN traffic to one of internet routing and intermediate node latency problems. Even duplex switching only adds bandwidth between each set of virtualized pairs or microsegmented LANs; communication between pairs or microsegmented LANs must still occur over the backbone, and as such, the bandwidth bottleneck is not necessarily resolved by this microsegmentation architecture unless you balance your segments properly and provide sufficient backbone bandwidth for the internet. A high-speed switching backbone between the microsegments is often necessary, as Chapter 5 describes.

Backbone interconnection

Creating a fast enterprise backbone for LAN interconnectivity does not always provide the extra bandwidth you seek. The traffic arrival timing and signal acquisition delay can cause significant latency that will waste the bandwidth of the backbone. This is true whether you install a wired backbone or a channel collapsed into a hub or router. Specifically, it is possible that several saturated Ethernets attached into the enterprise backbone will generate only 10 Mbits/s of backbone traffic due to the burst arrival packet distributions, buffer overruns, and outflow congestion, as Figure 3.20 illustrates.

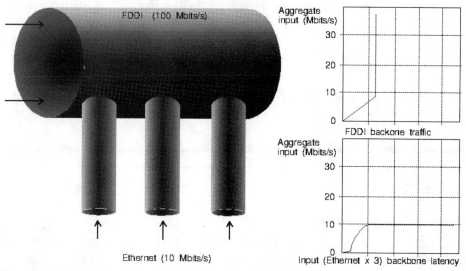

Figure 3.20 Adding bandwidth may not solve bandwidth limitations as the FDDI is limited by the Ethernet outflow.

This is an anomaly (when there is no packet buffering at the intersection) that you can model with many of the tools described in Chapter 4. It is a case of an exit ramp bottleneck at an intersection of two or more very fast highways. Hiding a problem with a faster protocol may incorrectly force you to consider a faster backbone, backplane, or WAN transmission channels. There may be cheaper and more effective alternatives for your particular environment and application traffic loads. Better you should model this than build your enterprise or think you will solve a bandwidth problem with a faster protocol. Replacing the LANs outlying to the enterprise with faster protocols will give a performance improvement for those LANs; the effective backbone traffic levels need to be modeled with an accurate view of interconnectivity method and devices before assuming the "solution" will work for your enterprise environment.

Let's focus on this issue some more because it really is important to understanding the nature of bottlenecks on the enterprise network; you cannot solve infrastructure bottlenecks by adding capacity until you can see the patterns in the traffic flows and identify the real constraints caused by chaos and turbulence. This "turbulence" is a feature of LANs and synchronous communications merging into the same infrastructure and has important ramifications in how you optimize performance. The infrastructure is not only the campus backbone, but also the entire collection of hubs and backplanes, router meshes, and service facilities. If you refer to Chapter 2 for the concept of self-similarity and the chaotic patterns that underlie network traffic, you will understand that merely adding the fastest components, fastest bandwidth, greatest bandwidth, and moving to the newest technology does not resolve the surface tensions in data traffic, the turbulence, or the eddies and whirlpools. Although turbulence is really evident in a data communication transmission channel, you will see it at intermediate nodes and backplane devices. Although Figure 3.21 shows a single channel (pipe) with analogous traffic slowed by internal turbulence, you might abstract this concept to waves and undertow on a curved beach or the complex convergence where many fast mountain streams merge.

Figure 3.21 Bandwidth and latency create eddies, whirlpools, friction, and transmission turbulence.

Protocol differences

Although most network operating systems typically use either TCP/IP or IPX/SPX, protocol differences create the most performance bottlenecks at the media level. Different transmission channels use different protocols; for example, ARCNET, IEEE 802.3 Ethernet, Quartet-signaling Ethernet, 4 Mbits/s Token-Ring, 16 Mbits/s Token-Ring, FDDI, ATM, frame relay, and modem links all have different packaging, cells, frames, packets, or XON/XOFF streams with different data payloads. See the NetSpecs database program on the CD-ROM for information about the many available data transmission protocols and data field or payload sizes.

Although NetWare tried to eradicate the packet differences by creating the NetWare Core Protocol (NCP) to transmit frames only as large as 512 bytes despite the native protocol packet maximums, the performance effects of frames smaller than the maximum packet sizes is actually significantly detrimental for all LAN protocols. This noble effort at creating a homogeneous NetWare frame has given way to burst mode protocol, large internet packets, and support for larger packets based on the protocol. Even multiple-hop packets are no longer limited to 512 bytes. However, the problem of translating or encapsulating one packet into another format is particularly acute on mixed-media enterprise networks. Although most multiprotocol router vendors claim translation at nearly wire-speed, the latency for delivery is about 500 μs for a frame translation. This latency creates a significant performance lag for multihop transmissions, and a traffic bottleneck if the device cannot keep up with the new arrivals.

Mixed protocols—including routable and non-routable ones—increase the potential for bottlenecks particularly at bridges, routers, gateways, and switches when decisions must be made based on the routing or destination address that is contained within the packet itself. When your enterprise network supports protocols such as SNA, TCP/IP, IPX/SPX, and DSLw, for example, you will probably need to both bridge and route the subnets and segments. Bridges are faster with mixed-protocol packets (if they actually handle the protocol), but routers perform better with TCP/IP packets and by necessity for heterogeneous protocol environments.[4] In situations where you have non-routable protocols such as LLC2 and NetBIOS, you will need to make provisions for encapsulating or bridging this traffic. Because of the widely acknowledged firewall segmentation and performance aspects of routing over bridging, encapsulating these protocols inside TCP/IP may be the best alternative.

The drawback to encapsulation is that it adds extra overhead and decreases available bandwidth, as the IP header is about 40 bytes. Worse still, encapsulating packets means extra processing for the router (to extract the destination address from within the media protocol packet, the TCP/IP packet, or within the SNA or SDLC transmission), so that performance slows considerably. The result is that many network managers end up turning off routing and reverting to a bridged network with devices that can forward the packets in their native protocols.

Latency is also of importance where SNA is concerned due to the encapsulation overhead. Data carried by SNA is often time-critical. LLC2 has timers that expire if acknowledgments aren't received in a preset amount of time; when the timer expires, the users' session with the host is dropped. Although DLSw promises to guarantee traffic throughput through routers to maintain SNA sessions over TCP/IP regardless of the link traffic level, and although early test reports show its reliability, hops and LAN traffic loads may invalidate this technology in your enterprise network. However, SNA would rarely time out in a local area environment due to bridge or routing latency alone. Lost sessions are attributed to accumulated congestion in bridges, routers, gateways, and switches.

Protocol fusion of LAN traffic with SNA includes SNA encapsulated into TPC/IP or IPX/SPX, and also TCP/IP and IPX/SPX encapsulated or translated into SNA. Fusion does work, but not well—LAN traffic on an SNA link fares poorly, usually due to the limited bandwidth and priorities assigned to the host traffic. Success with SNA on an enterprise link depends on network load, prioritization, and both the protocol fusion hardware and software. Some products work marginally; some do not work at all.

LAN traffic typically eats up all available SNA bandwidth allocated to it, and that may be insufficient. The requirements for synchronous SNA keep-alives and the burstiness of asynchronous LAN-type traffic do not work well together at moderate to high traffic levels. It is useful for occasional usage, but not for mission-critical connections, as the SNA sessions time out. However, just for the record, IBM tested SNA over an OS/2 Token-Ring 3174 gateway, with each workstation performing 6.2 transactions per minute and a total throughput of 938 (4 Mbits/s) and 1675 (16 Mbits/s) with under 20 ms response time. LAN channel

---

[4]Scott M. Bradner, Aiken Computing Labs, Harvard University, Cambridge, MA.

bandwidth utilization remained under 33 percent, although throughput was 1 Mbits/s; significant workload can be routed from a host to LAN workstations.[5]

SNA integration with LAN is filled with several other serious pitfalls. Technically, the SNA and TCP/IP (or other protocols) stack can either exist as a full protocol stack, or as a split stack where the host transmits with something that appears to be a standard FEP but is actually a dummied-up processor that then translates SNA into the native LAN protocol. The split-stack translation is apt to be slow. However, full-stack transmission may require that all hosts, servers, and clients have complete SNA and native protocol stacks active in memory. This can be an excessive burden, particularly for client operating systems with discontinuously-mapped memory, such as DOS and Windows.

Protocol conversions are not just a matter of repackaging the data contents in new frames. Internal network addresses, routing addresses, and checksums must be preserved as well. In fact, when the packet is encapsulated or disassembled, the routing device must read the native address so that the routing path and actual destination can be determined. This is difficult—actually impossible—to do at wire-speed. At excessive loading levels the typical result is network gridlock.

Multiprotocol stacks

The enterprise network usually integrates many media protocols (such as Ethernet and Token-Ring), two or more MAC-layer protocols (such as IPX and IP), and different types of routing protocols (such as boundary routing or RIP). Multiple protocols impose significance performance bottlenecks. Specifically, each protocol requires a special memory structure to compose packets and extract incoming messages, called a *protocol stack*. When multiple stacks are established for separate protocols on the same machine, this is called a *multiprotocol stack*.

One approach is to load the multiple protocol drivers concurrently into memory. This typically wastes an excessive amount of memory on PCs for DOS or Windows when such stacks are loaded for interoperability for any two or more NOSs such as Vines, NetWare, NFS, FTP, LAN Manager, and LAN Server. The traffic from multiple protocols usually is disproportionately higher when MAC and router protocols compete for bandwidth. Also the many protocols add load to all connectivity devices that must sort through them and reroute them correctly. However, most workstations and client equipment will ignore the extra protocols *without* any appreciable performance hit; they simply ignore packets not addressed to them, hence no extra load.

Network protocols

Despite many efforts to create other internetworking protocols or universalize driver communications with NDIS, ODI, and OSI protocol stacks, TCP/IP is the de facto standard for internetworking. Running TCP/IP on NetWare networks

---

[5]*LAN SNA Host Gateways Design for Throughput Performance and Availability,* IBM Corporation, ZZ81-0234-00, November 1989.

concurrently with IPX/SPX creates significant unnecessary traffic overhead. Specifically, the IPX window requires a one-for-one acknowledgment of packet receipt. Hence tunneling (encapsulating) TCP/IP within IPX/SPX generates more traffic than most network managers really want. In general, tunneling increases the broadcasts across an IP-based network. Other limitations include the double overhead on each workstation for the extra protocol stack, dubious access to devices on a native IP network, and the need to manage multiple address methods and map client address to IP addresses for each protocol.

Address mapping problems are more pronounced when routers are pressed into service to route and forward multiple protocols. First, there is usually a limitation of about 8000 addresses for all protocols in the routing table. Multiple addresses require that many more. There is the extra burden of handling the similar (but different) protocols. Protocol soup thickens when SNA, RIP, SAP, or the NetWare Multiprotocol Router travel the network channels. The routers can incorrectly redirect and repeat data. Also, routers base decisions on the NetWare address to filter by the IPX information, rather than the encapsulated IP address.

Slow FTP

FTP, the TCP/IP file transfer protocol, provides less than full media bandwidth at 2.52 Mbits/s.[6] A review of these numbers confirms that performance varies by platform, software implementation, and configuration and that performance up to 5.98 Mbits/s is possible on IBM RS/6000. Channel bandwidth is not the limiting factor; rather, software drivers and hardware efficiency are.

A disconnected device

Non-operational devices on an enterprise network create a burden out of proportion to their importance and their necessity. A "disconnected device" such as a bridge, router, or gateway creates a backlog for all traffic directed past it. If a disconnected device is reattached to the network, you are likely to see a blizzard of traffic related to prior requests and earlier time-outs. Furthermore, the effect of undelivered cells, frames, and packets creates a flood of repeated requests for service or repeated attempts to convey unsubstantiated deliveries. Furthermore, link failures usually precipitate a broadcast storm as other connectivity devices or service providers check access and revise or update routing tables. Sporadic and recurring link failures from a disconnected device also create a backlog even when spanning trees provide alternate paths, because the switching to the alternate paths is usually measured in seconds or minutes.

With mobile computing entering the consciousness of network operations, a disconnected device could also refer to a workstation powered down by a user on the road, or the station of a user on the road connected to the network via wireless cell technology or radio, and which is not generally accessible. Disconnected servers and workstations create a performance bottleneck, particularly when they

---

[6]*SNA vs. TCP/IP: Round One to SNA*, Kevin Tolly, *Data Communications*, Pp. 45-48, Jan. 1994.

contain embedded, or linked objects referenced by a master document. If you cannot access a key piece of information or even a small piece that is critical for updating or recalculating and recomposing a master document, you have a real bottleneck. Backup for remotely connected devices will increasingly pose a severe test of infrastructure performance; all those remote users will want to back up their data in case of laptop theft, power outage, or other road disasters.

Mobile computing also imposes another disconnected device bottleneck that is not as obvious as it might seem. While networks are usually designed and built to withstand a predetermined network traffic load for physically attached local and remote devices, few enterprise networks have been designed to date so as to include the performance effects of mobile, mostly disconnected devices when they *do* connect to the network. Because the interface for mobile computers is usually by cell-modem, the loading is mostly invisible until it reaches a threshold of traffic load that grabs the attention of network management. You will want to provide service for the ultimate day when all those disconnect devices are attached and active at one time. Also note that cell-transmission latency, including call-initiation, can impose significant setup times and multiple second latencies. See "Mobile and wireless" on page 65.

Ascertaining true causes for bottlenecks

Typical benchmarks used to stress networks, particularly the channels and secondarily the file servers, use block file transfers. The blocks may be as small as 32 bytes or as large as 4096 bytes. Because payload varies from 42 to 4962 bytes depending upon the transport protocol, you may see a transition between the block sizes and the data payload as actually carried by the cells, frames, and/or packets. As is often the case for benchmarks conducted on a cross-section of equipment for magazine articles, you will see that smaller blocks are more costly to move. The curve levels off at 256-byte blocks, as Figure 3.22 shows.

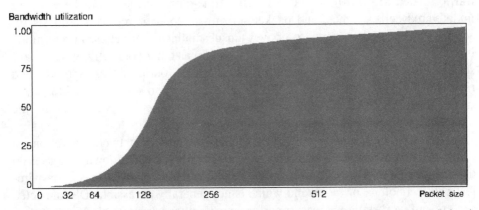

Figure 3.22 Repetitive block transfers increase bandwidth utilization for packet sizes larger than are normally found on enterprise networks.

Many formal benchmarks do not provide adequate assessment of performance as might occur within your enterprise environment. Most were developed as experimental concepts that have minimal correspondence to true traffic composition, routing patterns, or characteristics of LANs, much less enterprise networks. This is my opinion for the SPEC series, the BAPCo, and the TPC-A and TPC-B tests. (However, the BAPCo test, which emulates typical clerical workstation workloads, with runtime versions of Lotus 1-2-3, WinWord, dBase, and all the other necessary components is available from BAPCo for $99 on CD-ROM.) The TPC-C and TPC-D, which test transaction processing in complex environments, are complex and profound benchmarks that, unfortunately, are very hard to implement and to date have yet to be formally accepted by the TPC organization; we will not see implementations and trial results in quantity for some time. The ones used for "qualifying" vendor equipment and software are often based on sloppy or opinionated research. Refer to *Computer Performance Optimization* for additional information on benchmark techniques.

Reality for network performance varies from this primitive measurement. First, data payloads are rarely uniform. Second, a trace on network traffic will show a nonuniform distribution of payloads, usually distributed about a mean and usually skewed to the smaller packet sizes. Third, network load in a production environment is more likely to reach higher utilization levels for smaller data payloads—which are the norm anyway. What you are seeing is the effects of performance synchronization on the server and the effects of uniform outbound traffic. The task is run from the server, so while the disk loading is apt to approach 100 percent utilization, the CPU should be napping. If CPU utilization is the limiting factor for these tests and for your network environment, refer to Chapter 5 for optimization techniques.

Rightsizing and control of distributed systems

Although the shift from mainframe and centralized DP facilities to distributed LAN-based processing has reduced the costs for equivalent hardware and software and provided more flexibility and choices, few tools exist to manage the new enterprise and maintain security, integrity, and consistency. While this book is primarily about optimizing the performance of the enterprise network, it is important to realize you cannot deal with performance issues unless the basic day-to-day issues are reasonably in order. The order that you create defines how well you can optimize performance. You need the information to make informed and reasonable assessments; you will need management facilities for these functions:

- Software management
    - Software distribution
    - License management
- Configuration management
    - Infrastructure
    - Security (authorization and authentication)
    - Change control
    - Operations
    - Routing
    - Performance
- Information storage management
    - Distributed file systems
    - Backup and recovery
    - Tape library
    - Database administration
- Operations management
    - Enterprise management
    - Event management
    - Auditing
    - Troubleshooting
    - Print management
    - Performance
    - Planning and scheduling
    - Capacity planning
    - Accounting

You need information at all levels of the enterprise network infrastructure to make reasonable and informed decisions for optimization. The biggest bottleneck you might face is not the performance problems per se, but rather insufficient management and control information. For example, how can you determine the routing structure if you do not know what software users have, and what effect their load patterns have on LAN and enterprise latency and bandwidth? How can you assess security, reliability, and redundancy if you do not know the impact from LAN server backups of client workstations and the effects adding more such workstations? How can you assess security if you do not have a system controlling access to the enterprise resources?

Remote and modem communications

There is a huge vacuum for support of remote communications in terms of address planning, routing software, configuration, and remote troubleshooting. It is really new for most VARs, and support organizations. Because remote communications usually rely on modems, the foremost performance issue is both latency and bandwidth limitation caused by incompatibilities between brands and types. Check your actual transmission speeds and call setup times with a software break-out box that can analyze speed, error conditions, line quality, and other communications benchmarks.

When establishing remote modem (analog) connections at higher speeds, you want to optimize platform configuration so that it can actually use the full 9.6 Kbits/s (or faster) channel when available. By the way, most modem-based analog channels provide the maximum rated throughput unidirectionally at any one time. In other words, for example, 9.6 Kbits/s means that total communication in both directions (between the primary site and the remote location) is limited to that rate. Although telephone lines are connected with two pairs (four wires), in reality this *simplex* connection is based on one pair (two wires, usually a red and green). Do not generally assume that you can achieve 9.6 Kbits/s bidirectionally because the modem connection is *not duplex*. This will incorrectly overstate capacity two-fold.

If you apply compression (in hardware), you may get throughput at 18.6 Kbits/s, 28.8 Kbits/s, or even as high as 115.2 Kbits/s. However, remote network DOS or Windows nodes communicating at modem speeds will not be able to send more than 4.8 Kbits/s with remote access software when using the INT14 DOS interrupt. This is a fundamental software bottleneck. Also, the transmission speed is not usually a factor of the speed of the communications server. Typically, a Pentium will transmit no faster than an Intel 386. The difference is about 10 percent or less. The bottleneck is not the CPU, but rather the software driving the communications and the transmission channel capacity.

These problems are true regardless of whether the PC has the bug-free 16550 UART or not (see *Computer Performance Optimization*). It is a function of software, and you may want to try different remote access software to see whether it uses the INT14 interrupt. Notwithstanding, some remote access programs overrun the UART and serial port, and typically cannot achieve more than 20 Kbits/s with noncompressible streams or 40 Kbits/s with compressible data, even when the DTE speed itself supports a full 115.2 Kbits/s. PCs used as servers often have defective UART chips. Even with the correct 16550 series UART, PCs running Windows choke with 11,000 interrupts/s which is required for transmission rates of 115.2 Kbits/s (V.34 with compression). Windows on a Pentium usually cannot provide more than 1000 interrupts/s per port, which translates into 19.2 Kbits/s or less per port, or about 50 Kbits/s with compression. Throughput is also a function of data composition and burstiness.

Also, you want to learn how to work with the local exchange carriers. They can be an extraordinary bottleneck. Some carriers provide both private data lines and dedicated digital lines for long-distance communications. You may find that the only differences are hundreds of dollars per month, more expensive installation, and longer guaranteed response times to problems while latency and bandwidth are the same. This can be a technical and financial learning experience. Nonetheless, you know how to work with your local telephone companies.

As *Computer Performance Optimization* detailed, serial communications can be a bottleneck at the port; faster bandwidths for the telecommunication lines will not address these problems. UNIX is limited by port speed and throughput primarily by the CPU overhead associated with external communication. FEP support only so many channels.

Remote agents for clients and connectivity

Management of clients and connectivity devices represents a burdensome problem for large networks. Although it is easier to manage centralized resources because most management teams are housed near them, the enterprise network means that clients and the connectivity devices can be geographic distributed. Remote monitor, desktop management interface, and some proprietary software provide a remote agency for monitoring the health and performance of network workstations, bridges, routers, gateways, and switches. Use them if you can. SNMP and RMON support is probably a good consideration because this will help you integrate and standardize network management. However, realize that these protocols exact a channel bandwidth and device processing burden; see "Management overhead" on page 97 for more information about network management-induced bottlenecks.

There is also a more pressing reason for this capability from both an efficiency and performance perspective. Enterprise network bottlenecks are not usually located in the centralized facilities. They happen anywhere and a network panic can distribute the chaos through the network. Although you may think you have the cause and solution of a network bottleneck in front of you, you are likely to find that the cause is actually elsewhere. If you can monitor and manage that device remotely, you will resolve bottlenecks without your physical transit to location. Additionally, network panics create cascade failure. You certainly do not want to run around resetting devices to clear the gridlock while a new gridlock is being spawned just behind you as you proceed on the recently reset connectivity devices. Also, remote agents provide information in advance of real problems. You can set alerts and thresholds to inform you of impending problems. By the way, there are many beeper alert packages available so you can integrate the alarms with a real call to a person via the enterprise network, phone lines, beepers, or digital display pagers.

Limit access to resources

Bandwidth on a LAN is essentially free. The marginal cost is zero unless there is no spare capacity. When LAN traffic is routed over the enterprise, the cost is higher in terms of interconnectivity device loads; there is also a cost per traffic unit for the WAN links. You may also anticipate a very real operational cost on WAN links at capacity. Because it is easy for a user to make a mistake, click on a disk icon in a Window-based file manager, and export it to a machine across the world, you will want to manage access to the enterprise linkage. One other important WAN limitation is that addressing bandwidth limitations may take months of leg-work with your carrier. It can take as long as a month just to add another channel or increase the bandwidth after you have determined the newer requirements. Little mistakes like these are very expensive and threatening.

For example, it can easily take 18 working days to procure from the local exchange carrier (LEC) access lines. Long-distance interexchange carrier (IXC) service may require less time, but no less lead time. Problems are easily compounded if your service demands exceed the LEC's capacity to add service. While this may seem like a small-town problem caused by limited equipment or switching capacity at the central office, it is also happening in major metropolitan markets as phone numbers are running out.

Cost is not the real issue here—you can always charge it back and discourage users from making these mistakes—but rather a disk copy of any magnitude can bottleneck the enterprise. Although limited access to resources may refute the primary purpose of the enterprise network, it may provide the security, traffic prioritization, and management necessary for reliable and predictable operations.

Management overhead

The tools that you use to monitor traffic performance also impose a load on the network. SNMP and RMON, as well as workstation-based tools like the LANalyzer for Windows, or a server MONITOR (described in Chapter 4) come with a hefty price in terms of CPU requirements, RAM, disk space for storing statistics, and substantial buffer space to actually capture packets and packet information in real-time. Packet (cell and frame) capture tools and statistic compilers skew the performance of what you are trying to measure, because the measurement process itself requires substantial resources.

In fact, the more heavily burdened the managed device, the heavier the toll the management takes. This is an important concern for routers, switches, and complex gateways. Such a burden is the result of Heisenberg's Uncertainty Principle. Overhead for enterprise network monitoring ranges from 10 to 45 percent of total load for SNMP, but up to 100 percent for RMON and complex thresholds, alerts, or stream captures. Figure 3.23 shows what can happen when network management purloins the network capacity.

Voice over the network

Conferencing, paging, white board sharing on the network, and integration of telephony is a significant forthcoming topic. While many network managers and consultants think in terms of integrating data with voice, the reverse is also happening in that voice will also be integrated into the data networks. Voice, fax, data, on-line documents, and video will fundamentally be transported and processed in digital formats within the same infrastructure. Microsoft Telephony API (TAPI) is part of the Windows at Work integration and other efforts to provide unification.

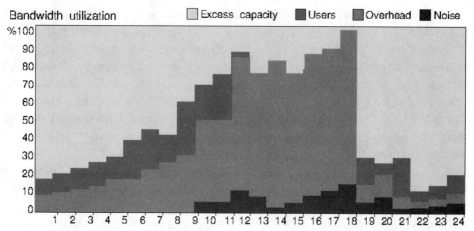

Figure 3.23 When management overhead is extreme, traffic can consist entirely of RMON and SNMP data punctuated by frantic threshold-exceeded alarms and alerts caused by the excessive RMON and SNMP data itself.

However, there is a difference between analog voice transmission and digital transmission. Sound is analog in general and can be transmitted at data rates equivalent to 4800 bits/s to 9600 bits/s. On the other hand, digitally encoded voice requires at least 17 Kbits/s in real-time each way. And, this is just for the sound itself. See Chapter 5 for methods of voice compression to alleviate this transmission bottleneck. Control of sessions, routing, paging, session enabling, and remote control will increase the overhead. When telephony is enabled through the network, you will also need storage for voice mail too. Store-and-forward, multiple calls, multiparty conference calls, and more complex services will require additional bandwidth. While 50 Kbits/s may not seem like much for a single virtual telephone connection, it is for WAN links; and particularly so when multiple calls are likely to be active at one time, for long stretches of time.

Imaging and video

Most video operations are going to be slow and represent a network bottleneck. Images, especially video in a digital format, are bandwidth-intensive. Although analog is more efficient, it has been effectively phased out by the demands of computerized and digital transmission. Even ATM cannot handle the load of digital video without data compression. An AVI movie with 15 frames/s at 200×200 pixels requires 14.4 Mbits/s. Voice over video requires another 17 to 44 Kbits/s depending on quality. Super VGA (XGA), with 30 frames/s at 1024×768 24-bit color, is 566 Mbits/s, and 5.66 Mbits/s even with 100:1 MPEG-2 compression. Even a typical VCR-quality material with MPEG-2 compression requires 1.3 Mbits/s which is faster than the 150 to 450 Kbits/s CD-ROM transfer rates and certainly a stress point for most PC-based workstations which cannot get more than 512 Kbits/s through an Ethernet adapter. When video and sound are integrated into real-time video teleconferencing, you will likely have a bandwidth bottleneck if this traffic is piped over the enterprise network. This technology may also be called data conferencing when combined with attendee access to a common computerized white board.

Fax imaging

While network-based fax servers typically handle most outgoing facsimiles well, most offer only a limited ability to deal with incoming ones. There are two issues, that of network load from the page images themselves, and that of the effort and overhead to determine how to route the file to its destination. Some of these limitations exist because there is no single standard for routing incoming faxes. Most network fax servers can receive, store and print incoming faxes, and let users manually route faxes to the intended recipients. The issue is the overhead for routing 100 MB files via the enterprise. See Chapter 5 for fax optimization.

Conclusion

Benchmarks and testing tools are critical for defining the important performance characteristics, gathering performance information, building relative and comparable benchmark data, and providing the means to make applicable performance evaluations. You will need to confirm your bottleneck hypotheses before embarking on the process of tuning a computer or network. Benchmarks provide the underlying information to pinpoint them. It will help you differentiate slow networks from slow client/server processing, a bus-bound server from a disk-bound client, and an overloaded server from overloaded clients.

If you are building a PERT or critical path model, the model will need path information, timings, and purposes. Those path values do not come from thin air. Furthermore, benchmarks provide the only effective means to assess the optimization effort. You will need to know if the computer system performance is actually better, that performance levels are going in the right direction—or for that matter, actually changing. Too often, optimization based on improper performance evaluations yields minimal or even negative effects.

Without a benchmark or measuring stick, you will be unable to gauge the effects of changes to your network. Chapter 4 now explains when to use benchmarks, which benchmarks to use, and how to gather information to locate performance bottlenecks. The tools include those that help you blueprint, design, analyze inventory, plan to cope with disaster, capture traffic statistics, and decode protocols. The next chapter demonstrates how to use *several* network simulators and determine throughput and latency capacity requirements against the available capacity. Also, the chapter applies these tools to compare bridges, routers, and switches in the enterprise network. Chapter 4 addresses which is the best interconnectivity device, a common question for network designers and anyone considering how to optimize internetwork performance and shows the surprising statistics of network load, packet sizes, and latencies that underscore the assertions made in this chapter.

Chapter
4

Tools

Introduction

Tools and models provide the only means to qualify the performance of the enterprise network, as well as the only means to model the performance of a new, modified, or redesigned enterprise *before* committing the resources and actually building it. In fact, you should have a good toolkit for all phases of the life cycle of the enterprise network. Unfortunately, there are no magic bullets specifically for performance optimization of the enterprise network, although a handful of hybrid and developing tools exist for tuning the GUI desktop, operating systems, configurations, and SQL or application code. The limitation to these tools are that they are specific to an environment, operating system, compiler, or vendor equipment. The most pressing limitation for the available tools is not that they can't be used to measure and model enterprise network performance in the same environment. Rather, they are often so difficult and time-consuming to learn how to use that they prevent the average network designer, manager, or consultant from realizing the benefits from these tools. You often need to know 100 percent of the functionality and ramifications to use them for measurement and modeling.

Nonetheless, you absolutely need tools to design, plan, test installation, test infrastructure, measure loading characteristics, manage operations, forecast traffic workloads and growth, and debug bottlenecks. Pick tools and learn their features, fallacies, limitations, and their true utility for your environment. Rudimentary knowledge of them—if you know even as much as 40 percent of the functions—places you at risk of misusing them. For those of you who do not have time to crack the manual—delegate the task to someone who will design careful models with fair data and interpret results well.

As a result, you need to acquire a toolbox of equipment and learn how to use general-purpose tools for the specific task of performance optimization. This chapter introduces this equipment and shows you how to use it for benchmarking and optimizing enterprise networks. It is important to realize that the enterprise is fundamentally a multidimensional entity with multiple routes, mixed protocols, many different paradigms for operations, convoluted software processes, and a wide range of computing platforms and connectivity devices; it is not the group LAN that can be optimized by the tactics of microsegmentation, protocol upgrades, and faster processors. You principally need to look at the intersections, but also at the connections and routes between them.

Performance tools are few

The enterprise network environment is so new, complex, and heterogeneous that comprehensive solutions do not exist. Although there are integrated network administration tools from a wide number of vendors, the need for solutions is so open that niche products appear and thrive. It is not clear whether the ultimate tools will ever exist either because technological change is presumed. The tools which optimize networks or operating systems are limited in what they do. There are basically three types of optimization tools.

There is the first generation of tools which do specific tasks. These have been available for the longest time. Disk defragmentation programs recover lost clusters, create contiguous disk space, and move frequently accessed files to the locations on the disks which are faster to access. The second generation tools are expert systems built on top of protocol analyzers. These have been available for about eight years and typically aggregate errors to make specific conclusions about the current state of LAN traffic performance. The third generation of tools try to quantify configurations which expert designers and troubleshooters make and automate this process. These three types of tools are discussed throughout this chapter. In overview, you will need tools for optimization which:

- Categorize process requirements
- Design network infrastructure
- Document and blueprint:
    Wiring routes
    Equipment
    Logical structure
    Process flows
- Inventory equipment and software
- Gather (capture) traffic statistics and analyze
- Optimize design and route
- Forecast

Design

The enterprise network should be designed proactively. It should not just happen or get cobbled together. Reality is, however, that most organizations clump LANs together, integrate mainframe operations, attempt to unify E-mail systems, cut host costs and downsize, and expand resources and services. Some disaster or event significant to upper management usually exposes the full extent of the communications chaos.

Blueprinting

Regardless of the heritage of the enterprise network—you will need to blueprint the infrastructure. There are at least five levels for mapping the network. This includes the geographical view, a detailed map of wiring routes, locations of equipment, a representation of the logical structure, and a diagram of process flows. These are useful for administrative and operational management; you will definitely need these blueprints for performance optimization. The following sections show the differences between these graphical illustrations, how you apply each type, and some of the tools useful for constructing these diagrams.

Geographical views

The enterprise network often spans a large geographical area and you need to document locations and connections. This is a high-level diagram to show all the facilities, services, sites, and routes. You typically need this information for the big picture to plan service levels, bandwidths, latencies, priorities, and the need for redundant or backup facilities. This diagram should not be overly-detailed as it is literally providing a birds-eye view. Figure 2.17 in Chapter 2 shows a provincial view of partial enterprise network sites in North America. You could generate this map with pages from topological maps, atlases, highway maps, and colored push pins. On the other hand, you might consider using a general-purpose mapping program, such as MapLinx or MapExpert, or using a general-purpose presentation program, such as CorelDraw used to construct Figure 2.17.

For the same reasons people use word processors instead of typewriters, consider computerizing this task. It is more efficient for handling changes, adds, and updates, and you can make unlimited copies. There are also special-purpose programs for data communications, such as NET•CAD or netViz, which is shown in Figure 4.1. The major advantages to these distinct tools is that the icons and items are specifically geared to diagramming data communications networks. Additionally, some of these tools provide zoom expansion to view site details to an arbitrarily minute detail. Some of these tools generate databases for reporting, administration, management, and performance analysis.

Wiring route map

While the previous illustration provides some indication of wiring routes, you actually will need more explicit information. Specifically, route information should include:

- Source
- Destination
- Carrier media
- Service carrier
- Committed information rate or bandwidth
- Connection protocol
- Latency
- Alternate channels
- Overflow congestion control method

The comprehensive clouds often used to depict X.25, frame relay, PBX lines, or ATM lack the detail necessary to indicate source, destination, matrix of connections, and the details you need. Since it also takes up a lot of the landscape, consider more pragmatic displays. However, the cloud represents a concept of multipoint access, and is useful for logical diagramming. You might note that a single transmission path may in fact comprise multiple paths, and protocols, as Figure 4.2 shows.

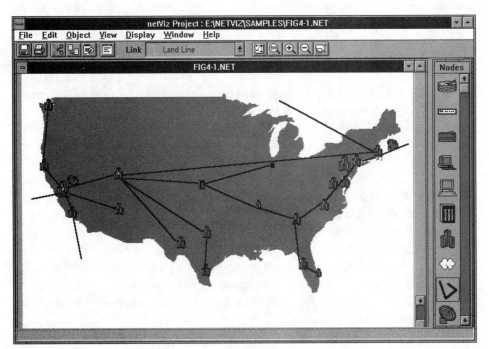

Figure 4.1 A geographical diagram of an enterprise network in netViz.

Although you think you know what the bandwidth is for a transmission channel, if that connection is important or expensive, you really need to test that channel to ascertain the true channel capacity. Most protocol analyzers—both for LANs and WANs—often include a traffic generator facility that will stress-test an existing segment. This will confirm channel bandwidth and its availability. However, sometimes you need to test performance and do not have access to the real resource. In those cases, you would dummy-up the channel with a device called a *link simulator*, as pictured in Figure 4.3.

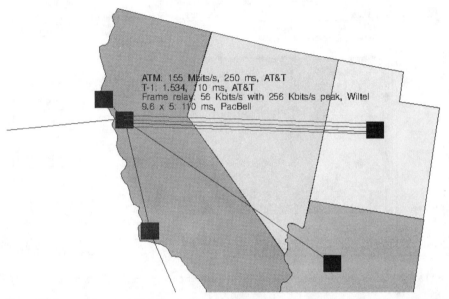

ATM: 155 Mbits/s, 250 ms, AT&T
T-1: 1.534, 10 ms, AT&T
Frame relay: 56 Kbits/s with 256 Kbits/s peak, Wiltel
9.6 x 5: 110 ms, PacBell

Figure 4.2 A connection needs definition in terms of what is actually in that connection, the bandwidths, speeds, latency, and committed information rates.

Figure 4.3 A data channel simulator (or link simulator) models a digital link for testing, developing, or demonstrating data and voice communication connections and performance effects.

Logical structure

A significant part of enterprise performance tuning is designing the network to meet requirements. You want to microsegment the network to increase the effective available bandwidth, but you also need to group the workload so that firewalls effectively contain most of the local segment traffic. You need to understand the structure of network, what processes occur on the enterprise and locally on the segments, but you also need to show what the equipment is, where it is, and (potentially) how it is cabled. One option is to create logical equipment placement on top of a site plan (scanned in as a bitmap or imported from a Windows application such as AutoCAD), as LAN•CAD illustrates in Figure 4.4.

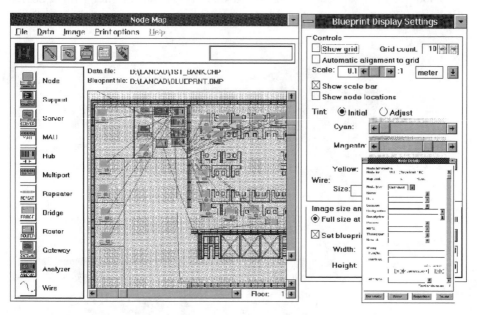

Figure 4.4 LAN•CAD shows relative logical placement of equipment with boxes or iconic representations.

A demonstration version of LAN•CAD, with a limited icon set of network devices and a count limitation, is included on the CD-ROM. Many tools are available for showing the network structure. Some people use PaintBrush because they have it and it is simple to use. MicroGraphix Designer and Corel Draw represent generic tools, though they are quite difficult and time-consuming to master. The growing ranks of more specialized tools integrate into the NOS environment or provide special features, much like LAN•CAD. I have included the demonstration (slide show) of GrafBASE on the CD-ROM so that you can be aware of your many options. One of the unique options for this tool is its integrated support for mapping and GIS as well as latitude and longitude coordinates for device placement, certainly a valid requirement for geographically distributed enterprise networks. The following shot in Figure 4.5 shows the drill-down capability and the specialized icons used for representing networks.

Equipment diagrams

Logical diagrams are great for design and intellectual performance tuning. However, you need to know where the equipment is actually located, and you need precise blueprints of sites, wiring conduits and paths, wiring closets, network devices, and user workstations. Logical structure doesn't directly show how units are physically connected together or what is actually connected without viewing the database or expanding the image. On the other hand, SysDraw can show exactly what is where and what the items actually are from their logos, as Figure 4.6 illustrates.

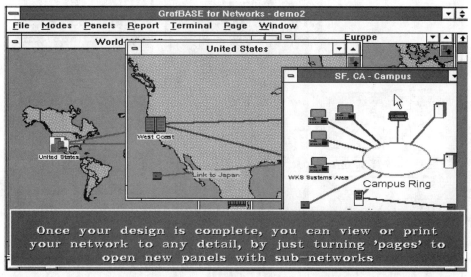

Figure 4.5 A screen shot from the GrafBASE demonstration included on the CD-ROM.

While SysDraw is sluggish to use, it provides full drawing functionally and snap to grid so that you can easily create an architectural floor plan with walls, furniture, and doorways, and show the exact vertical representation of the equipment. Note that the images are front view (sometimes back view for wiring connectivity), although most commercial blueprints are floor-view and show the walls, doorways, conduits, and building features. While SysDraw does not have detailed top views of equipment, you can use the other symbols to create realistic floor plans for equipment placement. The drawing accuracy for a particular vendor hub model, for example, is uncanny.

You can speed up SysDraw performance in part by turning the View/Show Fill off (so it is not checked), and View/Outline Dragging and View/Outline Combined Objects both on (so that it is checked), as shown in Figure 4.7.

Process flows
Accurate equipment images are one part of the documentation. It is important to diagram information and process flows. You certainly want to know about the network traffic, where it starts, what it does, and who or what is using the resources. While it is certainly possible to create fully functional flow charts with plastic templates and a pencil on paper, integration with order diagrams is

advisable. Additionally, process and information flows are not static and tend to change as the enterprise network matures; also, you are likely to discover hidden loads and tasks if you attach a protocol analyzer and decode the traffic. Refer to "Protocol analysis" on page 132 for examples of decoding protocols.

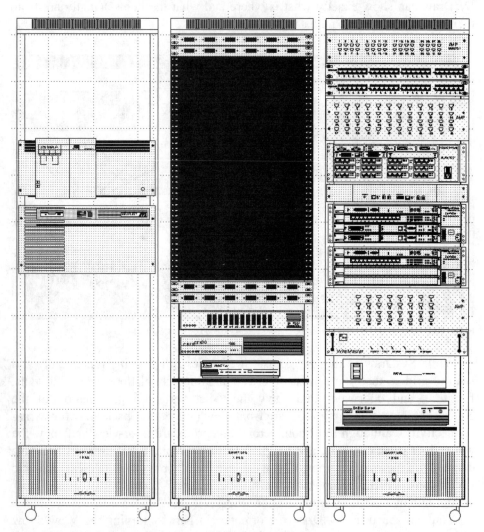

Figure 4.6 SysDraw displays communications equipment in exacting detail.

Enterprise network performance is adversely affected by an inferior wiring substructure. TIA/EIA has provided cabling recommendations—which are official or certifiably standards—for data communication wiring. Inferior wiring increases noise, crosstalk, signal degradation, signal impedance, and slow signal propagation. Although the backbone of the enterprise network is likely to provide adequate service, segments and subnets often were installed years ago and may not adhere to the necessary performance levels.

However, once you have the raw information it is useful to model it, and as Figure 4.8 shows, netViz can layer this information on top of the basic mapping and wiring models. The layers of information are explodable in what SysDraw technical support calls a "drill-down effect."

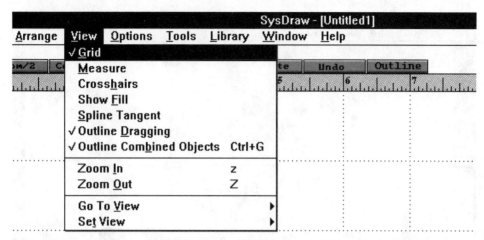

Figure 4.7 SysDraw performance is improved with three view options set as shown in this screen shot.

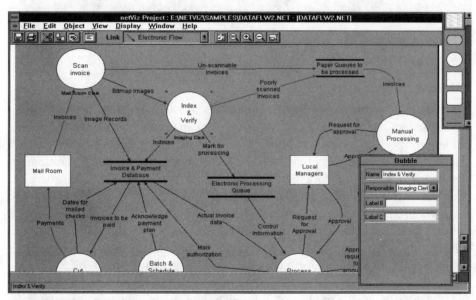

Figure 4.8 NetViz models process and information flows.

One of the limitations of illustration tools is that they represent a single-purpose solution to a multiple-requirement process. NetViz certainly provides more functionality specific to enterprise networking by incorporating the drill-down superimposition of network entities, but NetGuru and LANCAD additionally provide not only a drawing capability, but also bill-of-materials generation and network design qualification (with LANBuild) and simulation modeling of network performance (with LANModel). The illustration in Figure 4.9 shows the NetGuru Manager front-end creating the Token-Ring network schematic, which later can be used as input to the NetGuru simulation.

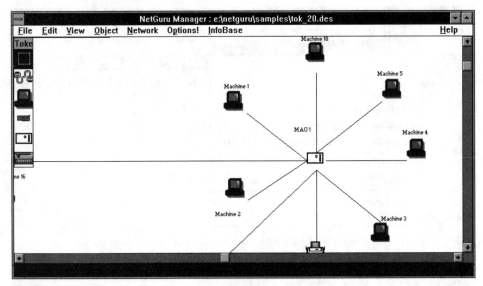

Figure 4.9 NetGuru schematic of a portion of a Token-Ring internet.

Inventory tracking

Knowing what is on the enterprise network is a continuing step in maintaining the equipment and site diagrams. Many network management programs (such as Cabletron Spectrum, HP OpenView, IBM NetView, and Novell NMS) have a discovery agent or process that can identify components and configurations of attached and active network devices. An SNMP-based monitor can query all other SNMP-compatible devices for identity information. The results are normally used to update the equipment diagrams, and if you can decode the routing path used to deliver these lists, you can use that to refresh the connectivity diagrams. SNMP is the standard cross-platform solution, because SNMP agents are available for more environments than any proprietary solutions and RMON extends data retrieval capabilities to all remote segments. UniCenter and OpenView are examples of administration tools that can correlate the information enterprise-wide. There is a number of tools available for specific platforms such as NetWare, Vines, DOS, and Windows (from Frye, Brightworks, Saber, Cheyenne, Magee, and Tally) with similar functions on a more restricted range of platforms.

In contrast, the information they supply is more pertinent to the typical PC-based environment. Some of these tools not only track basic information about the network stations, such as station type, but also more detailed statistics such as memory configuration, CPU type, CMOS configuration, PC adapter cards (by purpose and type), software on the local drives, and even revisions of boot parameters. Consider the value of these tools for locating shelfware—software requested, installed, and maintained, but never actually used—and as an effective means for managing and distributing upgrades throughout a large internet.

Although the quantity of information is perhaps more pertinent to asset management and auditing, some of these reports are pointedly useful for determining system-level configuration of user nodes. Also, because of the growing interest in integrating functionality and databases, some of these tools (such as BindView Network Control System) extend the standard network operating system information bindery or attributes databases. Secondary benefits of good network component and software inventory include financial control and monetary savings, better enterprise configuration, license usage control, and true asset management.

Spare parts

The enterprise network is a chain (hopefully with some redundancy built in) that is dependent on the functionality of all the interconnectivity and processing links. When a router, switch, or primary server fails, the cascading effect of device failure (while certainly profound to a segment) usually has a global effect on the enterprise. The most benign approach is reconfiguring around the failure. You might use hot routers, spanning trees, secondary paths, hot servers, or replication services. However, you cannot account for every eventuality, and sooner or later you will need spare parts. As the next chapter explains, standardization and homogeneity pay performance dividends; however, you still need to know what to maintain in a spare parts inventory. LANBuild (LAN Designer) primarily qualifies LAN design based on protocol limitations and generates bill-of-materials, as shown in Figure 4.10.

However, LANBuild also calculates spare parts needed as shown in the table on page 113. The spare parts are listed in the extreme right column. Notice that this parts inventory was generated for a mixed-media Ethernet network supporting both 10Base5 and 100Base-T. Critical parts also include cooling, circuits, and excess power to handle surges and failures of infrastructure, such as a Liebert 40 ton dehumidification and cooling tower or a 35 KW industrial building transformer. I have found that many enterprise networks include *legacy* LANs, which, because they expanded ad-hoc, outstrip the electrical and cooling infrastructure. This tool calculates not only the raw startup electrical loads that the network requires, but also individual circuit requirements as well. It also converts the load into waste BTU which must be removed for comfort and safety.

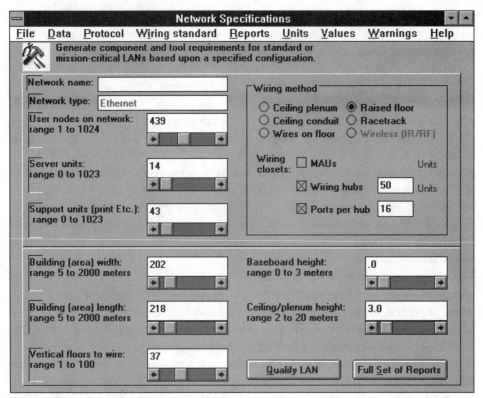

Figure 4.10 LANBuild (also LAN Designer) generates an inventory list of necessary spare parts based on statistical failure rates and relative importance of the parts to the continuation of the LAN.

Disaster recovery planning

The worst-case enterprise network performance is failure or downtime. Failure rates for major segments on the enterprise network are typically about 17 percent per week. This includes routers, routing tables, WAN linkages, servers, and other non-redundant components isolated with firewalls. Failure rates for the enterprise as infrastructure range from 3 to 10 percent on a weekly basis with downtime events ranging from 22 minutes to 6 hours per occurrence, depending on the research and consulting firm. Downtime costs the organization anywhere from $50 K/hr to $1.3 M/hr during this period. Downtime is a certainty.

It is not enough that there are many tools that provide asset management and fewer yet that help you track who is responsible for enterprise problems and how these are to be managed in the event of a crisis. A Disaster!Plan screen shown in Figure 4.11 highlights planning needs for operational continuity in advance.

The issue is not when or what will happen, but rather how you are prepared to respond to events in spite of whatever happens. Buildings are blown up, flood waters do rise 20 feet above the floodplain, and hurricane winds can remove equipment from inside office buildings.

A LANBuild Design and Planning Tool Report
Network Installation Bill of Materials
for Ethernet network

Date: 09-10-1993 10:43:12

Unit	Description	Quantity	Notes	Spares
Network	trunk cable (coax)	1		1
	The longest likely length is: 1412.00			
	Average likely length is: 743.00			
Network	KW electricity	216		65
Network	Dedicated circuits	302		91
Network	Tons BTU cooling capacity	192		68
Network	Hubs w/16 ports	10	2	2
Network	Trunk cables	51	2	6
	The longest likely length is: 426.00			
	Average likely length is: 154.60			
Segment	End connectors	4		2
Segment	Terminators	2		1
Segment	Terminator boots	2		1
Segment	Couplers (barrel connector)	1		1
Segment	Coupler boots	1		1
Node	Node workstation	439		44
Node	Server	14		8
Node	Support workstation	43		22
Node	NIC	496	3, 8	50
Node	Transceiver cables	496	2, 9	149
	The longest length is: 50 meters (limitation).			
	Average likely length is: 33.40			
Node	Conditioned AC power	107	2, 7	33
Node	Transceiver	50		6
Node	Tap (usually w/transceiver)	268		1
Node	AUI (or lobe) cable	546		55
Node	Cable ties	1660		
Node	Net user software license	496		
Node	Network software drivers	496		
Node	NIC device drivers	496		

Optional:

Node	UPS power supply	57	7	18
Node	Surge protectors	496	7	149

Structural Components: None included

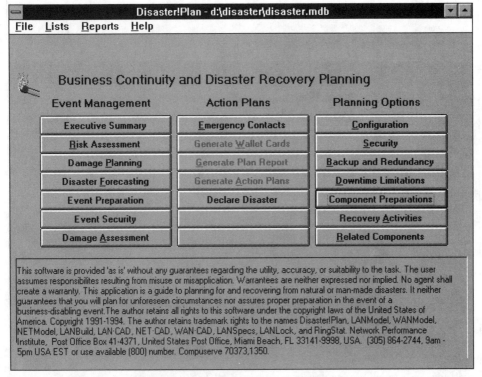

Figure 4.11 Disaster!Plan provides asset management and planning for organizational continuity in the event of an event that disables sites, connections between sites, or core business services.

While the backup tape manufacturers would have you believe that good image backups are "disaster recovery" planning and implementation, tapes that are faulty, tapes for equipment washed into the Gulf of Mexico or blown to bits by terrorists, or even tapes swept into the sky by a tornado are insufficient. A tornado in Oklahoma may suspend operations in Sussex U.K. because a local VSAT site has been knocked over, or a LEC substation lost electrical power for its SONET repeaters (that is why Chapter 5 illustrates the SONET ring with path protection). Key people die or are incapacitated. Vendors sustain their own disasters that may preclude your ability to get service or spares. You need plans for evacuation, succession, hot sites, remote network archiving, replication, and a plan of action that can be dusted off and immediately implemented for a generic catastrophe.

Figure 4.12 illustrates how a router is categorized for its importance or its potential to disrupt the enterprise operations, and the issues that require identification and planning to assure continuity in event that the device becomes inoperable for any reason.

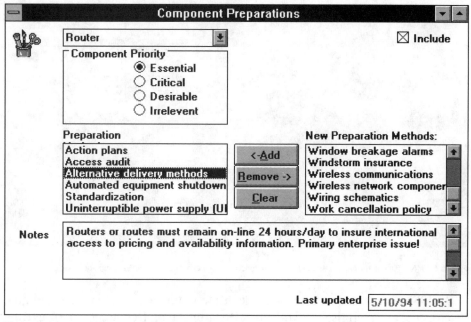

Figure 4.12 A router in Disaster!Plan is analyzed for its importance to the infrastructure and for the issues that make it essential for enterprise interconnectivity and operations.

Computer system and network performance measurement tools

It is useful to check the performance of nodes individually first, because it is easy and the tools are readily available. These tools are usually scripts or primitive expert system applications that codify the tricks that vendors and users have discovered over time. Examples include NetTune, Performance 2.1, or WinSense. Figure 4.13 illustrates how to change Novell server configuration with NetTune.

Anyone who has managed a network gets lost with the availability and effect of the many parameters. In a mixed server and NOS environment, this confusion is even more significant. Figure 4.14 illustrates more of the tuning handles specifically for NetWare, but nonetheless pertinent because analogous values exist for Vines, NTAS, OS/400, VMS, LAN Manager, LAN Server, and NFS.

The best use of these tools is to provide you with information in an on-line format that you might not glean from a user's manual, or to simplify the configuration of complex system settings. Performance improvements range from 2 to 6 percent. These tools can improve the individual performance of node, workstation or server, which can have profound effects on the enterprise bandwidth utilization and latency. For example, lack of file space on a server definitely slows file transfer performance.

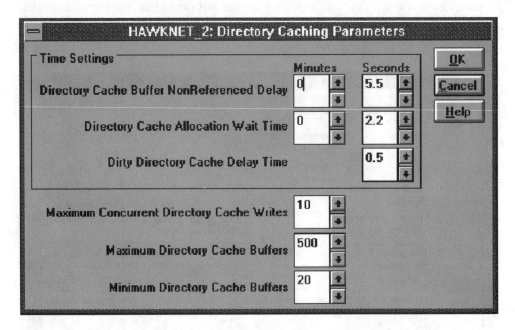

Figure 4.13 NetTune simplifies access to and optimization of the many esoteric NetWare configuration parameters.

In 1988, in an attempt to manage the different topologies and protocols in the heterogeneous enterprise network, the Department of Defense (DOD) and the commercial developers of TCP/IP released the Simple Network Management Protocol (SNMP). Since then, SNMP has grown into a widely accepted network management protocol. In 1990 SNMP became a TCP/IP standard, which further widened its acceptance and implementation. Basically, SNMP defines a database of network-management variables called a management information base (MIB), and the protocols used to exchange this information. It provides a common format for network devices and equipment, such as bridges, routers, gateways, and switches to exchange data via an agent with the network management station.

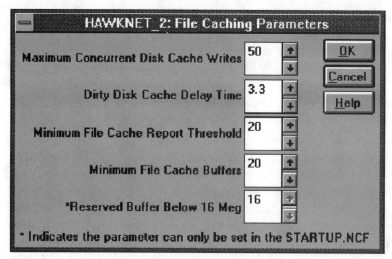

Figure 4.14 Other (more commonly recognized) NetWare parameters accessible through NetTune.

The agents are part of the software or hardware in the network connectivity devices. The *agent* typically consists of firmware and a specialized processor found in each SNMP-compatible device. The management software polls each agent over the network cable for pertinent information (using in-band transmission). The agent responds to the manager's request in a format designated by the particular vendor's implementation of the management protocol. The agents can also send link failures or alarms to the management stations.

The definition of SNMP includes using IP (Internet Protocol) for communications across the network, a carryover from the origins of SNMP as a management program run on the Internet; this can represent an overhead and performance problem, however, when the native enterprise protocol is IPX/SPX.

The more specific the tools or solutions are for an operating system or environment, the more likely that its impact will be noticeable and yield performance improvements. A few tools solve very specific bottlenecks, such as disk overloading, file fragmentation, insufficient configuration of disk cache, and sometimes how memory is used. There is a considerable number of memory optimization tools for managing DOS memory; these include MEMMAKER and QEMM OPTIMIZER. Use the SET commands and the NetWare MONITOR to determine if a file server has sufficient RAM. View the Tracked Resources display to make certain that the cache buffer pool, allocated buffer pool, and pool fragmentation is within normal ranges. Ranges above 50 percent free are normal. The message "Unable to load NLM" is another clear clue that memory fragmentation and insufficient RAM is hobbling the NetWare. Before concluding that insufficient RAM is the bottleneck, try rebooting the server to view tracked resources in a virgin state.

Analogous tools are available for optimizating the split between cache, RAM, and buffer structures for UNIX, NetWare, Vines, Windows, NTAS, and VMS. Some even watch traffic and disk utilization to configure the optimum size for disk clusters, sections, and disk cache. The most useful optimizing tools—this is true for hosts, minicomputers, servers, and client stations—are the disk file defragmentation tools; they can recover lost sectors and chain files into contiguous sectors for a 10 to 30 percent improvement in disk access speeds and transfer speeds. *Computer Performance Optimization* profiles them in depth.

Disk and file compression is generally a performance wash. The increased CPU time required to encode the files offsets the time saved reading and writing fewer blocks to the disk and sending compressed formats over the network. Perhaps when compression is included as part of the disk controller and the drivers are well-optimized, compression will provide the significant system improvements that are reasonable to expect. Note that compression used in conjunction with data transmission over the network does provide substantial effects, as Chapter 5 discusses with Btrieve data compression. However, because bandwidth is a fundamental limitation of the enterprise network, better integrated data storage and networking strategies are more likely to provide significant performance benefits. Data compression usually reduces files by about half. If the network load could be reduced by half, you would realize a significant reduction in enterprise transmission latency. Larger files, particularly data tables, images, and graphics, would benefit tremendously.

For example, RoboMon from CIS monitors VMS system performance by tracking and gathering file system and CPU statistics. Not only does this tool capture information, it provides an "advisory" service that recommends better operational parameters for each VAX. This is a more complex tool for an environment more complex than Skylight or WinSense, but one certainly that can help tune how the VAX interacts in enterprise environment running DECnet, TCP/IP, Leverage, or Pathworks (LAN Manager). Like many other network alert monitors, this will send pages via beepers or SkyTel messaging systems.

Vital Signs is a VTAM performance monitor for NCP connections, X.25, and SNA that tracks SMF, buffer pool displays, and a NetWare NLM interface. Vital Signs provides centralized management over LAN-based enterprise network connections for hosts providing client/server processing, file services, or host-to-network backup operations. The point is that some complex and heterogeneous management and data collection tools do exist for the enterprise environment. Although no SNMP consolidation is foreseeable in the near future, at least you can explore the value of some mixed-environment tools.

Stacker 4.0 also compresses files and ignored DOS disk sector boundaries. Although speed is not improved over baseline disk operations, space is re-claimed, with the net result that more disk space is available. The tools for monitoring file operations and disk utilization are more useful than native DOS or Windows tools. Figure 4.15 illustrates the Stacker monitor.

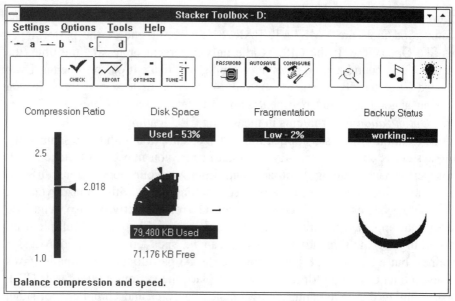

Figure 4.15 Stacker shows Windows workstation disk utilization in an easily interpreted format.

Although *Computer Performance Optimization* profiled the power of direct GDI printing technology (with Microsoft Windows Printing System or Laser-Master WinPrint, for example), it omitted the possibilities available with tools built into these and other applications. Many applications provide information for user debugging or technical support, or sometimes applications have debugging features and code that was never removed from the application. If it is there and useful, use it. For example, LaserMaster WinPrint provides a system memory bar graph that is at least as good as many other Windows utilities (Figure 4.16).

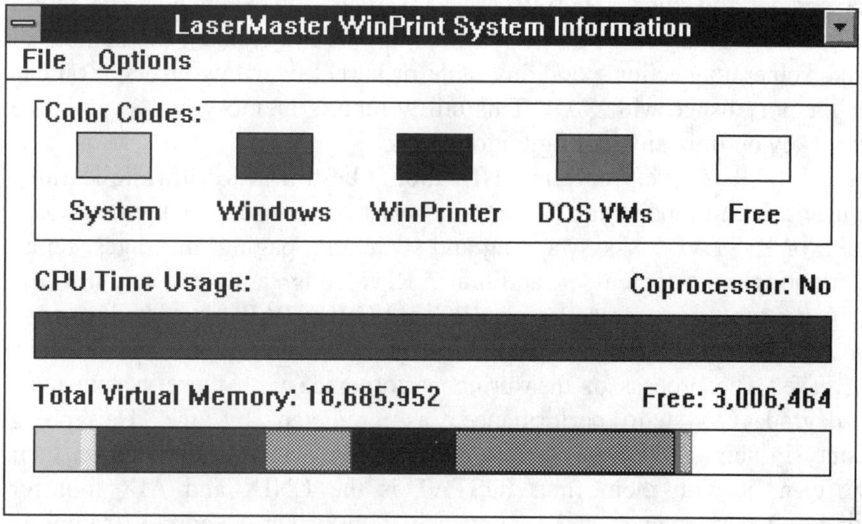

Figure 4.16 LaserMaster system information monitor.

Look at the overall distribution of CPU activity to get a high-level view of the system: good tools are VMSTAT, BAR, SYSUSE, or one of the performance tools like OpenView. If the CPU is idle only 5 percent or less of the time, the system is CPU-bound and no amount of tuning will make it perform better. Even when the CPU activity appears to be only 40 percent or less, make certain you are using a good tool and pertinent sample times. Refer to *Computer Performance Optimization* for problems in measuring CPU loading.

Also examine OS versus application CPU consumption with these same tools. If the system is consumed by API or system calls, you may need to take a top-down performance tuning approach and improve your applications' use of system calls. For example, replace old-style polling loops with event-driven I/O. CPU-bound systems bogged down by non-OS-related activity also send you directly into application tuning mode, as explained in Chapter 5. Multiple-user, multitasking, or multithreaded processes can be sped up with multiprocessor machines, but a single large job might need to be partitioned into smaller tasks (threaded) than can be multiprocessed. Or replace the system with a faster CPU. Also try tuning the code using a performance analyzer and compiler-tool suite. The benefits vary from an additional 5 to 10 percent performance gain all the way up to several-fold improvements if your application happens to contain code constructs that confuse even the best optimizing compilers.

Idle systems, on the other hand, are a symptom of an overloaded network or slow external I/O. Spreading the disk load or distributing input over more network interfaces will help reduce the time spent in I/O wait states. Finally, there is the problem of an idle system you wanted to be CPU-bound. Check the virtual-memory system paging rates, using VMSTAT or an SNMP agent. Look for excessive rates or time-outs. If you are running Open Desktop 2.0, TUNESH can provide very good insight into the kernel configuration. This is useful not only for servers and clients, but also for X servers and X clients. The report includes information about NFS file systems (import and export), the number of serial and Telnet connections, and any table or kernel overflow errors. You can monitor memory usage with SAR. This utility reports memory and swap space utilization, key optimization configuration items.

Although SAR is a very cryptic UNIX tool at best, it is worthwhile learning how to use it. It can report on CPU activity, I/O performance, background tasks (unlike PS or PSSTAT), task swapping and switching, paging, messages, semaphores, floating-point processor additions, RPC calls, and remote NFS file access. If you run this command with RPC and CRONTAB (described on page 121), you can build up a database of performance values on remote systems on a periodic basis. The process of measuring performance with these system tools alters (degrades) system performance, as predicted by the Heisenberg Uncertainty Principle. Therefore, users may not like your interference on their systems; clear it with them first. IOSTAT is the UNIX and AIX tool for measuring I/O performance and bottlenecks. It uses kernel address sampling performed every clock tick (10 ms) conveys TTY, CPU, and I/O subsystem activity. IOSTAT, which is similar to NETSTAT, is useful for comparing streams of asynchronous data transfers, CPU loading, CPU idle (100% - CPU loading), and the I/O wait percentage, which has the process identification description (PID) of 514 under AIX. IOSTAT does show you how long each process (by PID) must wait for either CPU or I/O resources. Although this is not a queue length measurement, nonetheless it does provide useful estimates for building the performance queueing model; multiply the process *CPU time* by the IOWAIT percentage to approximate the delay, and thus the queue length.

FSCK (with the -s option) is a UNIX file system defragmentation tool best run at system boot. It rebuilds the free data block list and cleans up fragmented files. Other useful tools include CRONTAB, which lets you run RCP or processes by time-clock. You can use this facility for load balancing, staggering process starts, alternating the server load (so that users think the system has better performance), and moving housekeeping tasks to off-peak hours. Most operating systems and NOSs have such facilities built-in or available as utility applications. Refer to the disk with *Computer Performance Optimization* for similar tools for DOS and Windows. If you are running SCO UNIX on an Intel platform, explore the u386mon shareware tool. TRACE and TRUSS will reveal NFS file access patterns in real-time. Shareware NFSWATCH generates NFS activity logs.

Windows NT and NTAS also provides a control panel feature for monitoring RAM and disk page swapping. Performance Monitor on NTAS also provides this information for the network environment as well. These services show when committed bytes of virtual memory approaches the committed limit setting and when pool non-paged bytes exceeds total virtual resources. Both are sure signs of server and workstation bottlenecks.

Novell provides its own MONITOR which not only measures packet traffic but also will track processor utilization. The trick here is to load the NetWare NLM with a "-p" command line option. This enables a menu option otherwise unavailable to capture and calculate CPU load, thread utilization, and interrupt overhead. TrendTrak and ServerTrak from Intrak provide these critical features for the NetWare environment. These management tools combine the functionality of LANalyzer for Windows and the server MONITOR with an integrated presentation and timeline views of CPU, cache, I/O, NIC and LAN activity. If you have experience in the mainframe or UNIX environments, you will implicitly understand the value of a better NetWare server monitoring. Because these tools provide information over time in a view not generally available from NetWare, there are demonstrations of these tools on the CD-ROM. Figure 4.17 shows several graphs generated by ServerTrak.

Tuxedo Enterprise Transaction Processing System (Novell) is a popular transaction processing monitor. It tracks TPS in most UNIX-based database systems, and monitors rollback and commit effects. Not only is this a tool for monitoring performance, Tuxedo can also cut SQL network traffic by optimizing queries with function calls. Most other environments, even SQL, often require a specialized performance monitoring tool for anything but primitive information. For example, Sybase SQL requires the SQL Monitor Server, which works only with Sybase SQL. SoftStar Job Scheduler tracks Oracle-based SQL jobs, determining how long each task requires and the effects on concurrent tasks. DBTune and DBAnalyzer from Information Systems Group provide Oracle-specific tools. SQL Watch from Pace monitors SQL record locking, which can create significant network delays when deadlock occurs. Ingres provides the interactive performance monitor for tracing record locking, memory allocation, and task performance.

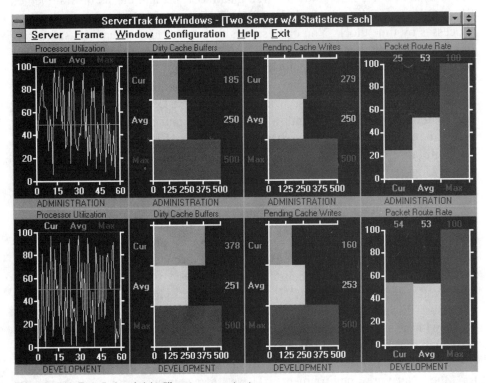

Figure 4.17 NetTrak display of eight different memory structures.

Do not overlook the native tools in the SQL environment, such as SQL Server I/O statistics or the built-in trace facility. Even the transaction log files may provide extraordinary information about what records and services are actually performed to buffer an ongoing process and then commit that transaction as a unit. This straightforward information actually may provide the most important verification of poor application design, what I feel is the most likely cause of remote access, DBMS, and client/server performance bottlenecks. If you are running the servers under a UNIX operating system, do not overlook the previously cited performance monitoring resources including IOSTAT, PS, PSSTAT, SAR, SYSDEF, VMSTAT, MONITOR, PERFMON (or PERFMETER), and PSTAT.

A user-level task may be bound by swap-disk speed because you're short of memory. You need to shuffle the job execution order, or work with developers to find a way to get their processes running in a smaller virtual-address space. Your local data center solves this problem with scheduling and workload management tools. Similar distributed transaction processing (DTP) monitors that are available for UNIX platforms include Computer Associates UniCenter, Open Vision's BatchScheduler, Auto Systems' AutoSys, Unison-Tymlabs' Load Balancer, and Aggregate Computing's NetShare and NetMake.

DB2 and the IBM environment has a collection of performance monitoring tools. These represent the front line for any DB2 performance tuning as outlined in Chapter 5. For example, the system management facility (SMF) reports internal activity for data records and can download this information via the enterprise network to a PC. Control block sampling (CBS) is an SNMP-like MIB facility that can be queried—after all, DB2 is a database—for activity information. CBS also allows for setting thresholds and alarms. The instrumentation facility interface (IFI) is an expansion of the CBS facility that relates groups and events to data activities. Furthermore, DB2 also has a built-in DB2 performance monitor, called the DB2PM. DB2PM reads the CBS and IFI traces to produce performance statistics about the environmental configurations, paging, database and table spaces, buffer pool utilization, checkpoints, user and usage anomalies, and record or file lock contention. The environmental descriptor manager (EDM) is a key for managing memory, threads, I/O, and memory pools. It also tracks DDl, BIND, COMMIT processing, PAGESET, DEAD-LOCK, TIMEOUT, CONCURRENCY, and log file activities.

Software metering

While initial network integration seems expensive, maintenance usually exceeds initial costs within two years. A significant portion of the ongoing costs is software and disk space management, and as more users are integrated into organizational networks the cost of generic software rises. Licenses are sold either on a per-unit basis, a site license, or for a fixed number of users per site. The expense for not having licensed software (when caught) usually exceeds a reasonable license cost. The problem of proliferating usage of generic software on the enterprise is actually significant for management and administration.

There are a number of tools (including SofTrack) that monitor numbers of copies of software on file servers across the network. These track who is using what software, and for how long. You do not want to overbuy or lock users from access to software that they need. Furthermore, excessive software typically creates a backup headache, which increases disk storage requirements, backup time and tape requirements, network loading, and control. You can also find out who has idle, iconized processes and what software actually is used. These so-called software metering tools can audit individual nodes on the network, monitor suites of applications, and even monitor the newer concept of add-in or plug-in tools, called *component software*, that integrate into suite applications such as Visio. Make sure that the tool you have can generate useful reports and graph utilization, provide cross-platform support certainly for OS/2, DOS, Windows, UNIX, and Macintosh, and reach across the enterprise linkages for full tracking coverage. If these tools integrate into the binderies, object databases, or other tools, so much the better. See also "Inventory tracking" on page 110.

Activity tracking

Computer Performance Optimization profiled two DOS-based tools called PC Sentry and FileStat, useful tools for tracking file usage and user system activity. Win, What, Where is a Windows-based alternative that tracks file access and user activity, and provides network tracking as well. It tracks how these resources are used and it also tracks idle time. See the CD-ROM for a demo version of this program. Figure 4.18 illustrates background tracking for a typical user's activity.

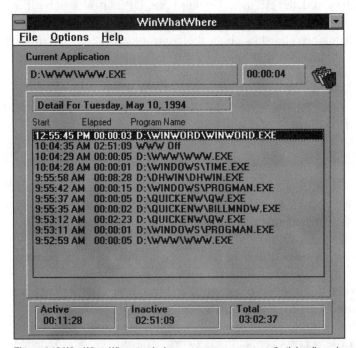

Figure 4.18 Win, What, Where tracks how a user uses resources (both locally and over the network).

This information is not only accumulated on a daily basis, but it is also saved for longer-term analysis. For example, Figure 4.19 shows the top activities and uses for a particular network user node. This aggregated data provides conformation for the types of resources and configurations optimal for different users. You can also use the interval of inactivity to make a valid justification for replacing systems with Energy Star-compliant systems, as Chapter 5 explains.

Disk waste

Neither DOS nor Windows has an easy method to find dead files. If you run UNIX, you can type at cshell:

find /home -type f -atime +30 - print> /usr/tmp/old30files

to locate all files not accessed (read or write) within the last 30 days. A direct enhancement of PC Sentry and FileStat is Disk Historian. This tool tracks all file usage without a DOS TSR and it can help you purge duplicate files from a user's workstation or files that are inactive. The information is particularly relevant for optimization enterprise network performance because you can track what is accessed from the server and what is local. Files that are duplicated on both the client station and server provide some freedom for making more disk space available. As *Computer Performance Optimization* detailed, free disk space provides room for file defragmentation, larger caching or temporary files, local swap files, and overall I/O performance improvements. Figure 4.20 shows tracking for a typical user's activity.

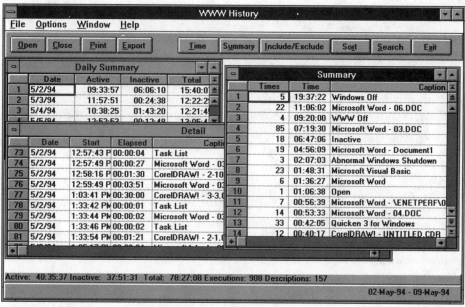

Figure 4.19 Win, What, Where also aggregates data and usage statistics over a long period of time, and provides some of the tools necessary to analyze the trends.

Additionally, Disk Historian provides detailed file information so that you can make intelligent decisions for disposing of inactive files and perform primary storage management, as Figure 4.21 shows.

Active data typically represents about 15 percent of data storage; inactive data, typically is about 50 percent of data storage, while the remainder represents configuration files and applications. On UNIX systems, about 10 percent of a disk represents the UNIX kernel, and another 10 percent is allocated for swap space. Windows demands about the same percentages for software and permanent swap space for optimal performance. Under Windows, Disk Historian can sort through 65 percent of a storage system that is data and identify the 80 percent of that storage that is dead weight. Even though on-line storage only costs about $1 per megabyte (but about $5/year to administer and back up), more sophisticated environments lean toward hierarchical storage management (HSM) systems to optimize costs and access to the data, while also providing a high degree of data integrity. As enterprise networks rival the importance of hosts environments, data storage management increases in its importance and effects on overall operations.

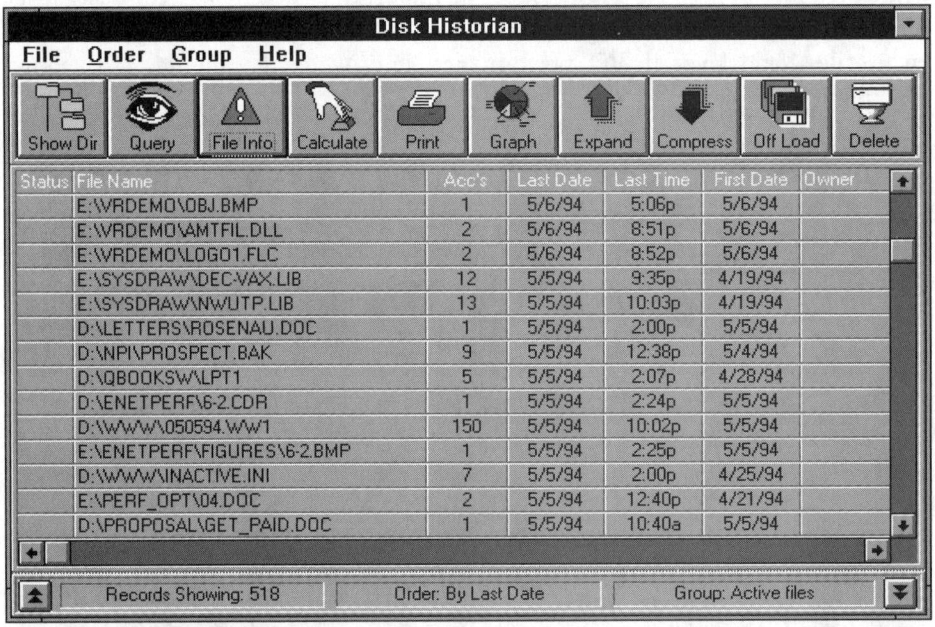

Figure 4.20 Disk Historian tracks daily and aggregate file access so that you can purge inactive files or replicate heavily accessed files from a server to a local drive.

Rummage

Disks frequently have multiple copies of the same files. Rummage for Windows searches your hard drive for duplicates based on file name, but also differentiates between CRC, file sizes, and versions. As you can see in Figure 4.22, software vendors do not always search for preexisting utilities. This is typical for applications in Visual Basic (VB), which has multiple versions of the primary .DLL files, but many .VBX software component files.

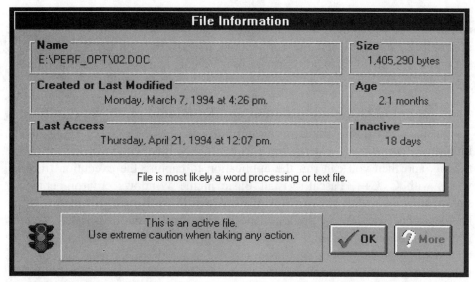

Figure 4.21 Disk Historian provides collateral information about selected files prior to compressing, archiving, or removing inactive, shelfware, or duplicated files.

Profilers

Because many enterprise networks are choked with software incompatible with the bandwidth limitations, routing latencies, and general resources, it seems wise to present some information on source code profiling. There are a number of tools available from many vendors, including Microsoft, Sun, HP, and IBM, that are geared to debugging, memory boundary checking, and optimization.

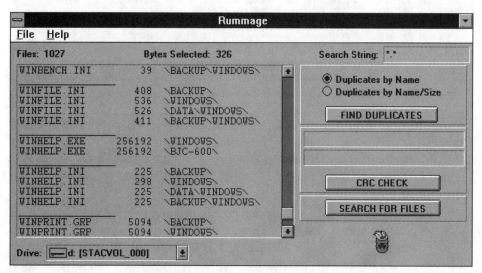

Figure 4.22 Rummage is a good supplement to Disk Historian because it actively searches for duplicate files.

For the most part, these tools are for programmers rather than for network administration and management. However, there is no distinction between design of enterprise networks and the applications that overburden them. You cannot view the network infrastructure as your only domain and assume that it will carry any load; even highways and city streets have weight, hazardous cargo, width, and height rules for traffic. You might rethink your network and quite literally apply some hazardous cargo rules. If so, monitor how the applications work.

One tool, called PinPoint from Avanti Software, fits the role better than most. PinPoint automatically inserts trace code into source code for performance tracing, and then you can run the application to capture the execution trace. It works with C, C++, Pascal, Visual Basic in the Windows environment, and should also work with most C/SQL dialects and PowerBuilder. I mention these languages because they are predominant for client/server and distributed network application development, and because I suggest methods for optimizing performance of these languages on the enterprise network in Chapter 5. PinPoint worked for me the second time I tried it on a Visual Basic application with MS and an Access database. The raw trace report is shown in Figure 4.23.

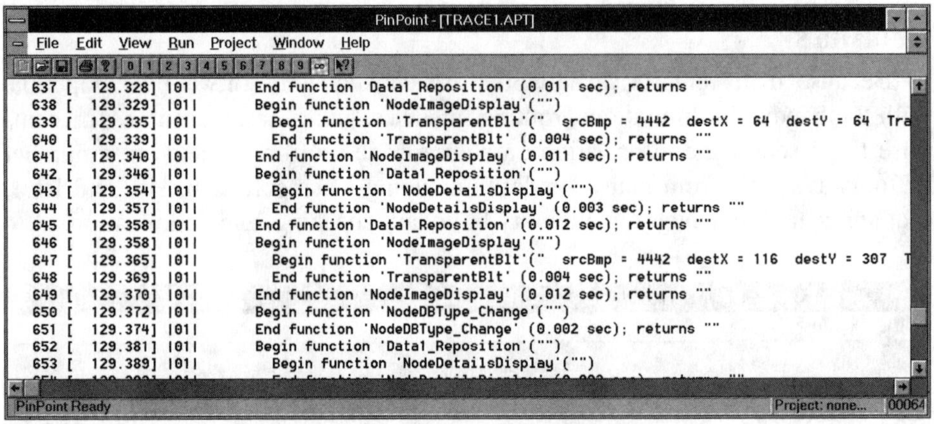

Figure 4.23 The raw PinPoint trace report on a Visual Basic application.

This part of PinPoint is not that much different from Pure Software's Quantify described in *Computer Performance Optimization*, or Microsoft's Source Code Profiler that comes with Visual C, SDK, CDK, or other toolkits. Figure 4.24 shows why PinPoint is useful to the network designer and management team.

For those familiar with object-oriented programming (OOPs), this profiler pinpointed the excessive use of change event triggers (instead of lost-focus triggers), which would be more efficient of CPU. PinPoint is different because it clocks events, executions, and process durations with microsecond execution times. This is very useful information for locating the application bottleneck.

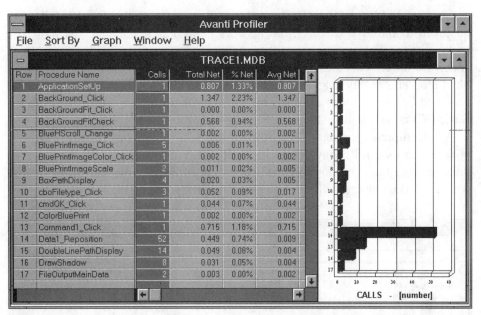

Figure 4.24 The PinPoint (Avanti) database and graphical report shows that some procedures (such as the Background_Click event) require elapsed time out of proportion to their frequency. Although frequent events have low overhead in a standalone environment (such as the Data1_Reposition) they are hogs on the enterprise network.

VB Compress from Whippleware marks unreferenced variables, duplicate variables, and will consolidate bloated code. It is more effective in attacking code bloat and some variable reference errors than a cross-reference variable index. Figure 4.25 shows the opening process screen.

Figure 4.25 The VB Compress processing screen.

Application designers forget the impact of routing, hops, and server access when building mission-critical software. As a result, they build stunning automation and information-integrating packages for a single user that fail when deployed to multiple users on the network. PinPoint is painless—as most test tools go—for ensuring successful upsizing and enterprise deployment. Dead code, fat, unreferenced code, duplicated routines, and rewrites of system calls bloat programs in any language. Just as LEX and YACC perform precompilation on C and C++ code, there are cross-reference tools from many vendors for Visual Basic (VB lacks a built-in facility).

VB Compress generates lengthy reports—nine pages on this small sample project—and the following screen in Figure 4.26 shows one small section. The information shows the unreferenced code and duplicate procedures, and shows what the VB Compress will do if you let it modify your code.

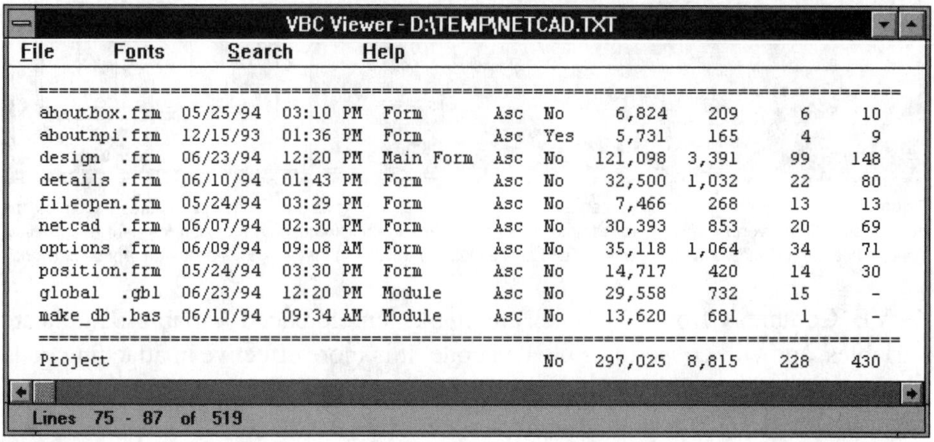

```
┌─────────────────────────────────────────────────────────────────────────────┐
│  ─ │                    VBC Viewer - D:\TEMP\NETCAD.TXT                  ▼ ▲  │
│ File       Fonts       Search       Help                                      │
│ ═══════════════════════════════════════════════════════════════════════════ │
│                                                                               │
│   aboutbox.frm   05/25/94   03:10 PM   Form        Asc   No      6,824    209      6     10 │
│   aboutnpi.frm   12/15/93   01:36 PM   Form        Asc   Yes     5,731    165      4      9 │
│   design  .frm   06/23/94   12:20 PM   Main Form   Asc   No    121,098  3,391     99    148 │
│   details .frm   06/10/94   01:43 PM   Form        Asc   No     32,500  1,032     22     80 │
│   fileopen.frm   05/24/94   03:29 PM   Form        Asc   No      7,466    268     13     13 │
│   netcad  .frm   06/07/94   02:30 PM   Form        Asc   No     30,393    853     20     69 │
│   options .frm   06/09/94   09:08 AM   Form        Asc   No     35,118  1,064     34     71 │
│   position.frm   05/24/94   03:30 PM   Form        Asc   No     14,717    420     14     30 │
│   global  .gbl   06/23/94   12:20 PM   Module      Asc   No     29,558    732     15      - │
│   make_db .bas   06/10/94   09:34 AM   Module      Asc   No     13,620    681      1      - │
│ ═══════════════════════════════════════════════════════════════════════════ │
│   Project                                          No    297,025  8,815    228    430 │
│ ◄ ┃                                                                        ► │
│  Lines  75 - 87  of  519                                                      │
└─────────────────────────────────────────────────────────────────────────────┘
```

Figure 4.26 A part of the VB Compress report.

Update management

Network managers have typically created stub programs on LAN nodes that would load an executable update program from the server. This process would occur on a timed daily basis or whenever the system was rebooted. The purpose for this was to provide a means to distribute configuration changes and new network drivers, update user software, or reflect differences in the network address tables or architecture. This saved a tremendous amount of time. When the LAN was simple, the scripts were simple. However, the time requirements for managing user nodes on enterprise networks with thousands of users and many types of platforms and network operating systems has made more efficient and automated update management more critical.

Specifically, performance improvements are typical with better (or correct) device drivers, client-side application changes, and bug fixes. What might be an excellent means to boost performance may not be feasible to install without the means to do it automatically over the entire enterprise. Additionally, some of the tools provide the updates as broadcasts, rather than as single repeated transmissions to each node on the enterprise network.

There is another side to inventory management as well. There is a financial cost associated with buying licenses for equipment that no longer exists, for users who no longer need it, or for retaining unused equipment in warehouses that are under lease or hefty month-to-month payments after lease termination. Additionally, software upgrades may require certain hardware configurations that transcend what is physically or financially available. These problems all represent inefficiencies, and in some cases actual performance problems resulting from inadequate network management.

Cable certification

Enterprise network performance is adversely affected by an inferior wiring substructure. TIA/EIA has provided cable recommendations—which are official or certifiable standards—for data communication wiring. Inferior wiring increases noise, crosstalk, signal degradation, signal impedance, and slow signal propagation. Although the backbone of the enterprise network is likely to provide adequate service, segments and subnets often were installed years ago and may not adhere to the necessary performance levels.

Although inferior subnet and segment wiring will degrade LAN performance when internet traffic is bridged, routed, or switched over these segments, the degradation will also affect performance enterprise-wide. You might want to rent, borrow, or buy a cable scanner, as pictured in Figure 4.27.

Figure 4.27 A typical handheld wire scanner. (Microtest)

The newer scanners also calculate TIA/EIA cable "certification" to levels as high as Category 5 (for 155 Mbits/s) for signal propagation speed, NEXT, signal isolation and capacitance. They are simple to use. You plug each segment, lobe, node wire, or jumper into the handheld scanner, set the wire type and performance parameters, and press the test button. The readouts typically show when the end connectors are installed correctly, if pairs are split, precisely where wires are crossed, whether there are breaks, shorts, or open circuits, and if the circuit is live. By the way, if you can purchase a unit with a hard-copy printer or connect it to a PC, you can create a useful document of baseline performance and a map of your wire infrastructure so that you can plan and implement appropriate upgrades to match performance requirements.

However, there is one caveat. Realize that the scanners only *qualify* the performance of the wiring to a recommendation; they do not *guarantee* wire performance or validate that your infrastructure will work with a particular protocol or at desired signal speed. Furthermore, these tools will not test that pushing transmission speed beyond rating will not cause radio emissions that will aggravate neighbors or startle the FCC. Nevertheless, these tools are useful to locate the 80 percent or so performance problems that are wiring-, connector-, or patch panel-related. The scanners are also useful for qualifying whether IBM Type 1 (for Token-Ring) might support copper-based FDDI (CDDI) or ATM. These tools are invaluable for discounting physical reasons for poor performance.

Protocol analysis

LAN Performance Optimization details the use of protocol analyzers for LANs. The use of protocol analyzers on enterprise networks is not much different. They work the same way, although you want to capture data for each segment individually and also capture the traffic sourced from one segment, destined to a different segment, and which other segments that cross-traffic also affects. See "Enterprise network cross-product traffic" on page 171 for more information about protocol and traffic analysis for the enterprise

The CD ROM provides a demo of Novell LANalyzer for Windows (LZFW), which can run on any Windows platform and with many network cards. Version 2.1 is available for UNIX, IBM networks, NFS, and SNA with remote agent support. This addresses some of my concerns for the limited NOS and environment support with prior versions. You want the same LAN-type statistics at a minimum, which LANalyzer provides, but accumulated for subnets and segments:

- Bandwidth utilization (throughput)
- Latency
- Error rates
- Breakdown of traffic by purpose, protocol, source, and destination

The following screen shots illustrate the use and utility of LANalyzer for Windows. Figure 4.28 illustrates an overload situation. Figure 4.29 illustrates the excellent on-line debugging help available with the NetWare Expert feature and the Windows help facility.

You will need distributions over a time period, not just incident values. Additionally, you will need such statistics for all segments and backbones (or backplanes). Every segment, LAN, or simple pipe between pairs of devices represents a network subnet with the potential for performance bottlenecks, as Figure 4.30 illustrates.

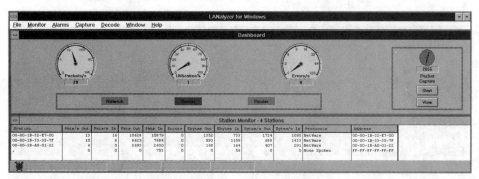

Figure 4.28 LANalyzer for Windows showing a network overloaded with traffic.

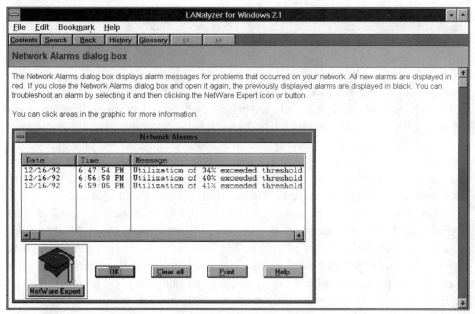

Figure 4.29 LANalyzer for Windows on-line optimization and debugging help.

Intel's LANDesk is another interesting NetWare management tool and protocol analyzer that provides one feature particularly that is worthwhile to me. Its application monitor is a promiscuous mode packet capture routine that decodes every packet by application. In other words, you can track network usage by application—whether it is the client/server accounting package, WinWord, mail, database applications, or even games. It even provides trend analysis. Although any astute network management-type who knows how to use a protocol analyzer should be able to get the same results with Sniffer, Azure, or any other analyzer for that matter, the LANDesk software and graphs simplify the process; it becomes a rational task to monitor enterprise workload and begin a long-term load planning and balancing project. A side benefit of the application monitoring facility is for locating shelfware and other marginally-used software.

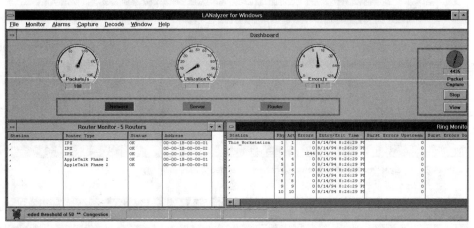

Figure 4.30 Capture statistics for all segments and interconnecting pipes on the enterprise network.

SAS Institute offers another alternative for capacity planning and resource management with SAS/CPE for VMS, MVS, and open systems, such as UNIX and UNIX derivitives. These tools are primarily data collection and capture tools for network traffic and download data from SNMP, Spectrum, and other protocol analyzers. However, they are capture disk, MONITOR, SPM, and accounting information as well. It is important to realize that the SAS software is not the first line for protocol analysis as it depends on hardware or software agents for source data. Instead, you apply the SAS statistical tools for advanced analysis.

What does differ is the need for monitoring of remote segments, distant LANs, and WAN connections. By the way, what makes the LANalyzer for Windows particularly useful is that at $1500/unit, it is inexpensive enough to provide for each NetWare workgroup, load onto a few laptops, or for network device I/O pairing, as explained in "two protocol analyzers" on page 135. There are other inexpensive software-only protocol analysis tools that are not NetWare-specific, including Azure. However, most LAN monitors do not have the capability to provide WAN cell and stream decodes, although some vendors sell multiprotocol portable decodes for LAN and WAN traffic. SNMP is useful because it supports RMON, alerts, alarms, and thresholds. However, recall that the inherent limitation of any in-band management is that network bottlenecks can prevent timely delivery of information.

RMON is useful because it uses remote probes beyond the view of most protocol analyzers. The agents often retain traffic and performance data on the device until specifically polled rather than broadcasting across the enterprise network for performance reasons. RMON devices can be polled to extract historical information. There are nine groups of MIB objects:

- Statistics—measures probe-collected statistics such as the number and sizes of packets, broadcasts and collisions.
- History—record periodic statistical samples over time that can be used for trend analysis.
- Alarms—compare statistical samples with preset thresholds, generating alarms when a particular threshold is crossed.
- Host—maintain statistics of the hosts on the network, including the media access control addresses of the active hosts.
- HostTopN—provide reports that are sorted by host table statistics, indicating which hosts are at the top of the list for a particular statistic. Note that host does not mean mainframe but any RMON server.
- Matrix—store statistics in a traffic matrix regarding conversations between host pairs.
- Filter—allow packets to be matched according to a filter equation.
- Packet capture—allow packets to be captured when they match a particular filter or threshold value.
- Event—control the generation and notification of events, which may also include the use of SNMP trap messages.

The Token-Ring RMON MIB contains four extensions that deal exclusively with Token-Ring-related functions:

- RingStation—gather statistics and status information for each station on the local ring.
- RingStationOrder—provide information about the monitored stations, listed in downstream ring-order sequence.
- RingStationConfig—collect station information, such as a station's MAC address, and its assigned physical location.
- SourceRoutingStats—obtain source routing statistics, such as the number of frames designated as a single-route broadcast.

RMON comes with a hefty price in terms of CPU requirements, RAM, disk space for storing statistics, and substantial buffer space to actually capture packets and packet information in real-time. Note that RMON (as with SNMP) does skew the performance of what you are trying to measure because the measurement process requires substantial resources.

Also, you will probably need *two* protocol analyzers for enterprise network analysis. As Chapter 5 explains, node and virtual switching make it possible to put a monitor on any segment almost instantly—thus making it less necessary for each subnet to have its own analyzer. However, you still need two units to track the incoming traffic to an intermediate node versus outgoing traffic to validate the node performance; the paired units let you see what actually passes through a connection, and by the process of elimination, what traffic is actually filtered or dropped by the intermediate node. This paired technique is invaluable to test the routing performance of a router, switch, hub, or other device which appears to be losing traffic, quenching overloads unnecessarily, or misrouting packets. If you are testing complex spanning trees or paths with multiple routes, you may need a monitor for each possible route. Figure 4.31 shows how you would use multiple analyzers to test the throughput performance of a device suspected to be performing improperly.

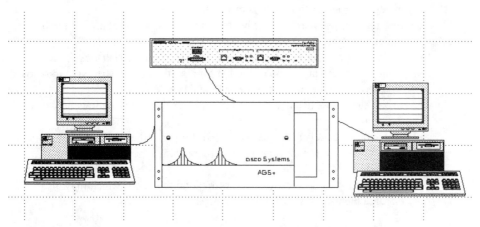

Figure 4.31 Two analyzers are used to test the performance (reliability) of a routing device.

Be careful in your blind acceptance of RMON. Although this protocol is supposed to be compliant across platforms and vendors for accessing SNMP statistics, this is not always the case. Data retrieved from the MIBs may be inconsistent, inaccurate, truncated for lack of available storage space, or in error because counters were not properly reset. Also, RMON agents and proxy agents—simplified and smaller applications for running in DOS or Windows environments in place of memory- and resource-hogging full-blown agents run on routers, servers, hosts, and devices with lots of cache and RAM—do not always return a full set or complete trace of device activity. Captured statistics may not be accurate, current, or complete.

SNMP, OSI CMIP, and RMON are not the only acronyms related to network management and operations. You might also include X/Open, COSE, CMG, OSF, and OMG. The Internet Engineering Task Force (IETF) is the group charged with standardizing SNMP technology worldwide also is involved with management frameworks. Its Host Resources MIB extends SNMP to PCs, servers, and UNIX hosts, and is part of the IBM Systems Monitor/6000 MIB. SNMP-2 (or Party MIB) allows one system to support multiple agents that can monitor the host and the enterprise network simultaneously. Because vendors double-talk standards, you need to know what is best for the enterprise network and what standards relate to SNMP, SNMP-2, and CMIP, or what are vendor extensions without widespread support. The next few paragraphs outline the most commonly referenced organizations and networking standards.

X/Open Company is based in Reading, U.K., and is the most influential of the UNIX-oriented standards groups. The nonprofit corporation has approved a range of documents since 1984, including the X/Open Portability Guide (XPG), which details the requirements of vendor independent operating systems used in distributed nets. X/Open working groups meet bimonthly. Ongoing work includes management of UNIX hosts on LANs and internetworks. Some working group members are writing APIs geared specifically to systems management. These will be added to X/Open's approved Management Protocols (API-MPAPI), which is used to create applications that support SNMP and CMIP. The XMP API has been incorporated by Bull HN Information Systems Inc., IBM, and other vendors as part of their frameworks.

Common Open Software Environment (COSE) represents a consortium founded in 1992 by prominent (non-Sun) vendors. COSE is defining a set of specifications to be deployed by all UNIX vendors in their products. These will establish a common graphical user interface, object-oriented technology, multi-media, and systems management. Specifications are based on existing technologies, but modified for vendor-independence to work with different UNIX platforms. The group is writing APIs for user group security administration, print spooling, backup and restore (and HSM), software license management, and software distribution. COSE is also defining ways these services might be represented as objects in request brokers based on the OMG's Corba. COSE also seeks to eliminate differences in UNIX-based utilities used in systems management applications.

Object Management Group, Inc. (OMG) is based in Framingham, MA. It also is a nonprofit corporation of vendors, users, and developers looking to promote object-oriented technology. OMG's most significant achievements was the 1991 release of a Common Object Request Broker Architecture (Corba) which is a set of APIs for creating an Object request broker (ORB) that accepts data requests from various applications and matches them to mechanisms that retrieve information. The OMG is defining a directory service for multiple ORBS that will allow data exchange, pertinent for OOPs on enterprise networks. Other projects seek to define specific objects that represent management information.

The Open Software Foundation Inc. (OSF) is in Cambridge, MA. It has designed the Distributed Computing Environment (DCE) establishing an open operating system which can run on workstations from leading vendors. Also, the Distributed Management Environment (DME) is a set of services for DCE, including software distribution, license management, event reporting, and subsystem management. Political infighting among DME supporters has resulted in a series of missed release dates and led to the OSF's decision to jettison its own object request broker. OSF will add APIs for use with Corba-compliant ORBs, a mechanism called the instrumentation request broker. Despite OSF's troubles, vendors of management platforms say they plan to make their frameworks compatible with DCE/DME, and the technologies used by OSF are finding their way into other standards efforts.

The Internetwork Management Forum (IMF) is an international vendor/user consortium dedicated to establishing systems and net management guidelines, and is based in Bernardsville, N.J. It has defined various network components that can be used in object-oriented frameworks, such as those based on Corba and CMIP. IMF is integrating a set of APIs, called the open management edge, with the OSF instrumentation request broker to bring mainframes and other legacy systems under the control of object-oriented frameworks on the enterprise net.

A major limitation of switched virtual circuits is that each circuit consists of only two nodes at a time. Therefore, it is absurdly difficult to capture protocol information with a protocol analysis tool. You cannot eavesdrop or listen in to the conversation without building a subnet (with a bridge, router, or gateway). You are limited, given the current crop of monitoring software and hardware, to the native protocol monitoring tools built into the switch. However, the cisco Catalyst is a switched port analyzer for RMON MIB data gathering, troubleshooting, and extended internet tuning.

How to apply protocol analysis
The general utility of enterprise network protocol analysis is:

- Locate duplicate addresses
- Track resources that cannot be accessed
- Isolate nodes that cannot boot or access server
- Isolate nodes or subnets that cannot communicate with others
- Trace the path of lost E-mail or incomplete deliveries
- Track excessive service advertisements (SAP requests or optimal path searches)
- Monitor dropped sessions or host time-outs
- Watch incomplete file transfers
- Trace excessive router redirects
- Validate the load from broadcast storms
- Test for slow response time (long latencies)
- Display excessive bandwidth utilization
- Display excessive collisions (Ethernet) or token losses (FDDI or Token-Ring)
- Isolate line and burst errors
- Locate the source of ring errors

Duplicate addresses are a particular problem when LANs are first integrated into the enterprise as network managers tend to assign the simplest names and numbers. It is the simplest approach that runs afoul of the addressing needs in distributed networks. Typically, search the RIP tables, Yellow Pages, or ENS directories for suspicious entries, and then decode addresses to confirm this common problem. Limited access, lost E-mail, and failures to boot or communicate represent a mismatch in addresses or device and domain names. Typically, you will need to decode packet addresses, routing addresses, and the progression of traffic from source to intended destination to unravel this operational gridlock. SNMP-compliant devices may obviate the need for multiple protocol analyzers because you can extract data from remote MIBs.

Dropped sessions and time-outs indicate excessive latencies, faulty media, device resets, network overloads, bandwidth overloads, or configuration errors with hosts, servers, or clients. It is easier to increase session timers, and that may solve the problem. Look for expired activity timers, too. On the other hand, you should use SNMP or an analyzer to time-stamp packets at the source and the destination (if it even gets there) to construct transmission latencies. Check bandwidth utilization—that's easy—for overutilization. You would probably know that beforehand, though.

Protocol analysis is the tool to spot excessive traffic, SAP traffic jams, broadcast storms, errors, line and burst errors, and traffic overflows. The packet source (or destination) is usually indicative of how to solve this problem. In fact, you cannot miss it. Otherwise, capture data and parse it by protocol and type. It helps here to maintain a history log so that you can differentiate endemic problems from acute ones. Endemic problems indicate infrastructure failures; qualify the wiring, NICs, connectors, and panels with a scanner.

A routing loop is a special problem for the enterprise network because there are likely to be quite a few routers; some networks have hundreds. Review the alternative loops, the routing protocols in place, and check the accuracy of the routing tables themselves. You are more likely to find that SMNP-managed routers or unified router management software simplifies this protocol analysis task, as opposed to extracting traffic with multiple hardware- or software-based protocol analyzers. The alternative requires progressively tracking packet streams across the network until you locate the loops or split (and recombining) paths.

Management statistics

LAN Performance Optimization provided basic techniques for using the protocol analyzer. You might want to refer to that book, or several other available titles specific to NetWare, Vines, NFS, or the protocol analyzer that you might own. Although monitoring traffic on the enterprise network is not too much different from tracking LANs, you need to remember that there are likely to be multiple protocols, multiple routes, local traffic destined for remote sites (cross-product traffic), a higher level of background traffic related to network monitoring and administration, automated configuration, remote monitoring, backup, timed events, messaging, E-mail, and remote access. The primary parameters include:

- Traffic loads (bandwidth utilization)
- Routes
- Errors and overruns
- Thresholds and alerts
- Paging

- Latency
- Composition of workload
- Bottlenecks
- Frame sizes

Parameters useful to the client/server portion of the enterprise network include:

- Application transaction response time
- Server CPU load
- Disk latency and throughput
- Records read/written per second
- Application transaction response time as a function of transaction load
- Application transaction response time as a function of the number of users

Types of benchmarks

A benchmark is a form of advertising that provides results-oriented promotion based on some particular computing configurations. Make certain you understand the guidelines for the benchmark, how it can be applied, and what techniques can be used to skew results. There are no benchmarks designed to specifically test the performance of the enterprise network. Many are "interesting" but not accurate or indicative of performance on your network.[1] Typically, benchmarks stress some performance aspect over another, or try to be fair and create separate results for different components in the system. However, just as computer systems are interrelated, the enterprise network is an infrastructure and cannot be truly assessed based on the performance of its individual components.

One of the serious disadvantages of benchmarks is that vendors pick and choose which benchmarks they make available for their systems and devices. Many (but not all) game against the community for building equipment that is primarily faster on the standard benchmarks. In fact, the OEM and vendor attitude toward benchmarks can be best summed up by an article referenced with the table of contents byline, "Playing the benchmark game can be profitable."[2] However, the article does not present such a nefarious view as the copyrighters created. A key point of the article is that hardware or software benchmarks represent an abstraction from the true operational environment and are often meaningless. However, by abstraction as well, any benchmark that characterizes your operating and enterprise network environment realistically is a useful tool.

No two enterprise networks look alike, share the same equipment, have the same traffic patterns, have the same carriers, or have the same routes and connectivity devices. Although benchmarks exist for NFS, TPS loads, block

---

[1]*Benchmarks: How Accurate Are They?* Jake Richter, *Cadence*, pp. 35-37, April 1993.
[2]*Endgame*, Paul E. Schindler Jr., *OEM Magazine*, p. 88, April 1994.

loading and file transfers, they typically characterize a utopian or simplistic workload. Scott Bradner at Aiken Laborations (Harvard University) has tested routers with 30-second bursts of traffic at fractional rates of the theoretical capacity of the protocol. He has used 50 percent rates to test throughput, full bandwidth rates to test packet loss and forwarding rates, and small time-stamped bursts to test latency. Although this technique provides a good generalized indication strictly of router performance, it does not provide much information in the way of internetwork performance or the processing infrastructure. For example, it does not show how routers will perform with unbalanced bi- or multidirectional loads, and it does not address filtering, firewall, or multihop performance issues.

Even more complex system benchmarks (such as TPC-A, TPC-B, Dhrystones, Whetstones, MIPS, and some of the SPEC, which are profiled in *Computer Performance Optimization* and included on the tools disk with the book), do not reflect the environment of large networks, only the systems themselves. The system benchmarks are useful to qualify the performance of key servers because they often represent a substantial bottleneck for the enterprise. BAPCO end-user application scripts and IBM client/server TPS profiles are interesting, but not representative of your enterprise. HINT, designed by the Scalable Computing Laboratory of the U.S. Department of Energy, is a supplement to the older SLALOM benchmark and is easier to use. HINT measures the amount of work a computer can perform over any range of time, where the work is defined in units called QUIPS (quality improvement per second) or MQUIPS (millions of QUIPS). The workload characteristic is based on the integration of the function:

$$\int_0^1 (1 - x) / (1 + x)$$

Frankly, this type of workload is uncharacteristic of most business and clerical computer tasks. It is neither representative of network bandwidth or CPU intensity. The labs at *Client/Server Today* have expressed that the SPEC benchmarks are best used to compare compilers and are undermined by the KAP preprocessor and optimizer. As a result, they have created Clue as an enhancement of the previously-cited SLALOM benchmark, and altered the function as shown:

$$\int_0^1 4 / (1 + x^2)$$

I cannot see why this is any better than the original. Nonetheless, SLALOM, HINT, and Clue are scalable in duration and problem size to resolve short interval problems with SPEC benchmarks, tests parallel SMP-type servers, and does work well on client/server systems. It is useful for exploring the effectiveness of the cache, the cache size, client paging, and hence network I/O. Nevertheless, do not base client/server architecture or network design decisions on a benchmark with a workload so far removed from your environment. Use it only to stress-test the configuration and the system and network architecture.

Although TPC-C and TPC-D promise a more realistic workload characterization, few database vendors actually have working code because they are much more complex than the financial bank posting transactions represented by TPC-A and TPC-B. Needless to say, these benchmarking tools are quantitative in that they will show what performs better within the limitations strictly defined by that tool. For example, PERFORM2 will show how well the server can perform disk block-oriented I/O, and if the networks can sustain the packet rate; it will not provide a qualitative measurement of how good your network is, how much it can be optimized, or what should be done to improve performance. Even as I write this, TPC-E for measuring on-line transaction processing (OLTP) on enterprise networks is in development, as is TPC Client/Server for OLTP measurements in (simple) client/server environments. By the way, neither of these two benchmarks measures the effects of OLTP and the network performance separately even when you are testing OTLP client/server within the environment of the enterprise network. This only goes to show that the enterprise infrastructure is complex. The SPEC committee is also developing the SDM (System Development Multitasking) and SFS (System File Server) and the SPEC I/O suites. (By the way, SFS is really nothing more than a new name for LADDIS.) These benchmarks are designed to characterize workload from multiple users (nodes), but only on a single system. This also underscores that the enterprise infrastructure is complex.

The best benchmarks far and away come from tracking peak loads on your enterprise for all channels and devices (with SNMP, CPU loading traces, operating system monitors, or a protocol analyzer); add this information to your blueprints and network diagrams. Such numbers represent real values for your environment and you can use this information when you use a modeling tool. Be certain that you match these peaks with some performance basis, if at all possible, so that you can establish thresholds and performance service levels and extrapolate design and infrastructure changes (subject to the restriction that network loading levels are nonlinear). However, it is important that you benchmark a steady-state environment. If you test the enterprise or a client/server process when it is first activated, you are likely to get false performance readings that include one-time initiation overhead, as shown in Figure 4.32.

Realize that the data that you collect today is irreproducible. Historical data is very difficult to recreate. It is useful to retain old results and the test configurations so that future comparisons can be made, and differences in environment, workloads, platforms, and methodologies can be factored from the historical data for comparisons with the changed environment.

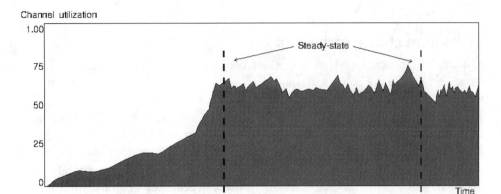

Figure 4.32 Benchmarks should be performed during a steady-state for accurate results.

While *Computer Performance Optimization* included many benchmarks, it did not profile X Windows tests. Two are mentioned here for reference only and do not necessarily replicate the workload for your enterprise network. See also "Workload characterization" on page 144. Xmark, X11Perf, and Xbench are benchmarks for X clients and X servers running X Windows. These are popular X benchmarks that calculate an Xstones performance indicator. X11Perf is slightly more popular and sophisticated. Xmark from X Performance Characterization group (XPC) is geared to X server performance and is a reinterpretation of X11Perf. Given its industry source, it is unlikely to reflect your environment.

Applying benchmarks

Network organizations will be wasting time if the use of the benchmarks does not have prior commitment from upper management, and if recommendations will not lead to network design or performance changes. Weed out those who aren't serious about change, as they undermine the process. Primarily, you want to determine in advance what different results will mean, and create guidelines for interpretation so as to remove the element of politics and turf wars later:

- Benchmarkers should be clear from the start about what information they are willing to share and what they're not.
- Determine as early as possible the issues and processes to explore, making sure they're in sync with your organization's needs. (Avoid hidden agendas.)
- Maintain a healthy skepticism about the numbers. There is a temptation to exaggerate results or hide weaknesses.
- Beware of benchmarks drawn exclusively from one industry or vendor.
- Run benchmarks during peaks or off-peak to verify loading and basis infrastructure response times.
- Start with some common way of evaluating processes and results, as you want to know ahead of time what to expect and what conclusions you can possibly make with the benchmark results.

Workload characterization

Nonetheless, benchmarks are important because they provide a simple tool to stress-test the enterprise network and get a general feeling for how much capacity it has and where the bottlenecks might be. It is still important to realize that PERFORM2 or PERFORM3 will probably overload the server I/O before anything else, whereas a client/server imaging application with data compression is more likely to overload the CPU first. The front line tool for workload characterization is you; it isn't some tool that promises to answer all your questions about applications, network traffic levels, user response times, and record locking overlaps. Instead, question users about what they typically do with their computer. Work logs or transaction processing logs are good sources for gathering statistics about client/server or database operations. If your environment does not currently maintain transaction logs, assess the performance hit for activating them. Because many database applications maintain transaction logs in case it is necessary to rollback, restore lost work, or synchronize replication servers, you may only need to archive the logs rather than to recycle disk space.

When the enterprise environment provides less centralized application services and more white-collar office work, Disk Historian and Win, What, Where provide the logs necessary for assessing the network workload. (The networked version of these tools track file and application access by network user.) The workload can also be captured and parsed with a protocol analyzer, such as LANalyzer for Windows. It may require considerable effort for someone to categorize the workload by task, but it is feasible. If the traffic is chaotic but varies little from day to day, you can rebroadcast (some percentage of) the captured traffic in addition to the normal workload traffic. This provides a means to stress the network with the same type of load you envision increasing.

Note, however, that rebroadcasting capture traffic has two problems. First, it is direct to network nodes already creating the same traffic load and approximately the same composition. You are likely to overload the clients when that is not really the test you envisioned. Second, LAN traffic is based on messages, and old messages can create unexpected results as clients reject them or servers respond that the master records already exist, files have been deleted and do not exist, or relational child records do not fit within the current state of the database. You can bypass these limitations by rolling back the state of the database and moving all the users (from the captured sessions) to equipment with new passwords and authentications. Also, testing tools that capture a session as a user typed it or vary a session with random passes through the menus also can provide a characteristic workload.

Tuning tools

Hawknet NetTune is a tool that helps you set the server configuration parameters for various releases of NetWare. It is much like Performance 2.1 for OS/2 as described in *Computer Performance Optimization*, although it does not offer suggestions for what the settings should be, or what settings will improve performance. It does provide a simplified front-end to the more complex task of adjusting the .INI configuration settings in an editor, or using the SYSCON commands and the multitiered menu system.

Performance 2.1 provides more direct and windowed access to SET parameters in SYSCON, and other .INI files. It does not suggest better values or provide performance suggestions. You will need to watch MONITOR or use the LANalyzer for Windows (see "Protocol analysis" on page 132) or a similar tool to see the results of the setting changes. Note that some parameters may have no immediate effects on the NetWare server until the system is rebooted. If you are confused by the NetWare SET parameters, this tool defines the purpose for each parameter as well as the default and acceptable range. It provides the same confusing on-line definition as Novell's but at the additional cost of an NLM in memory on the server.

Modeling the enterprise network

Many people misunderstand the difference between *measuring* network performance and *modeling* network performance. They incorrectly assume that they are the same. Measurement is the process of gathering performance and operational statistics on a system as it exists. All the tools presented thus far in this chapter capture workstation, server, device, channel, or gross enterprise network performance information. This information is the performance measurement of how the measured item performed in the present and the past.

This information is useful for locating existing bottlenecks, wiring failures, overloads, implementation problems, and likely places for new bottlenecks to occur with changes to the design or increases in the workloads. It does not show you how the network will perform in the future unless the future looks exactly like the past. Modeling is the process of constructing a representation of a system as existed, exists, or *might* exist. If you need to predict performance in the future, you need to model the future network, although you usually model with prior measurements. Typically, you model enterprise network performance to determine *network capacity*, *channel bandwidth capacity*, and *latency*. Network capacity relates to the infrastructure as an entity, whereas the channel bandwidth capacity relates to the transmitted traffic volume obtainable through a pipe, path, intermediate node, server, or host. The latency refers to not only individual buffer and service times at specific devices, but the aggregate point-to-point service time. Additionally, while latency is a primary indicator of transmission times, it is also a secondary indicator for where a path is critical or slack; as

Chapter 3 explained, latency is a highly useful tool to indicate global minimum and maximum values. The next sections show you how to determine network capacity, channel capacity, and latency with:

- Extrapolation
- Simulation
- Statistical modeling
- Emulation modeling

Extrapolation modeling

Extrapolation is the process of extending current knowledge. It is theory, conjecture, prediction, and guesswork. Most attempts at extrapolation are typically linear extensions of preexisting information. For example, consider a network with 60 users where bandwidth utilization is currently at 60 percent and response time is at 1 second (great!). Because this yields one percent bandwidth utilization per user, the maximum sustainable capacity is exactly 40 *more* users on the network. This ignores overhead, management, interprocess, and timing effects. It ignores the effects on response time. This may work for SNA or asynchronous connections and simple LANs, though usually not (particularly for Ethernet). It will not work for enterprise networks. Traffic loads and latencies are typically nonlinear. Okay, you say, let's gather more information to create a better prediction. Lower workload on the network to 30 users, note that bandwidth is now 15 percent and response time is now 0.25 seconds. I plotted a linear prediction based on those two points and also created two nonlinear estimates through those points, as shown in Figure 4.33.

Clearly, a nonlinear extrapolation model is more representative of backlogs, traffic loads, and how bridges, routers, and gateways actually function. Even if you gather data for other points, the question becomes which curve equation not only fits best to all the data points, but also best predicts beyond the range of the test data. Refer to the previous figure. Almost no experienced network person would accept the 100 user capacity. But, would you bet your job on enterprise network performance capacity of either 65 or 80 users? Would you keep your job if you stated that 60 users—the current level—was the maximum sustainable capacity? You need better tools to accurately establish that capacity and performance curve. Extrapolation modeling is not very reliable, and cannot predict bandwidth utilization and latency—the crucial enterprise network performance measurements—simultaneously.

If you can find an environment identical to yours, you can use that as a benchmark and extrapolate that environment more effectively to your own environment. Most often, though, your enterprise network will be different enough to need better performance modeling tools. The necessary capacity planning tools are based on object simulation and statistical models.

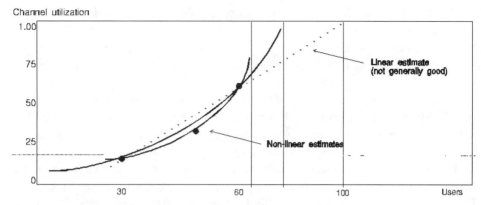

Figure 4.33 Plotting bandwidth and extrapolating capacity based on prior information.

Simulation modeling

Simulation is the logical process of imitating physical processes. Like extrapolation, simulation is based on prior data, as is all network performance modeling ultimately. However, simulation modeling requires the construction of a more complex model than extrapolations. It is based on the behavior of tasks, processes, devices, and interactions. There are two fundamental types of simulation modeling: discrete and state-change.

Both discrete and state-change modeling apply prior data to define the behavior of these tasks, processes, devices, and interactions as object templates. These behavioral templates are then used to create a representation of the network you want to model, whether it already exists or has yet to be built. The logical objects simulate the activity of real-world components and events, such as bridges, routers, transmission channels, broadcast storms, transactions, servers, and workstations. Simulation is different from linear or even nonlinear extrapolation, because you are combining the extrapolated behaviors of many objects into a larger framework that often provides nonlinear results. Almost always, enterprise network performance is a nonlinear function of network device counts, traffic levels, and throughputs.

The discrete process modeling is a step-by-step series of equations into which the prior data, either raw or massaged into mathematical distributions, is fed so as to create the bandwidth and latency predictions. Only processes that can be defined (and are defined in the model) are actually included. The state-change modeling method uses a different technology, called Monte Carlo simulation, to create a set space of network conditions and events that can alter the network conditions. Refer to *LAN Performance Optimization* for an explanation of states and state transitions. Relevant states include, for example, router buffer full, router off-line, no traffic or low traffic volumes, and especially steady-state conditions: events are traffic loads, failures, and application processes. Only those states and events that are defined are included in the model. Factories and production processes have typically applied discrete process modeling, whereas state-change models have primarily been the province of statisticians.

With either method you are not limited to modeling preexisting conditions, as with extrapolation models. Rather, you are extrapolating the behavior of devices and processes and combining these into larger designs for real or imagined networks. Simulation does not limit you to extensions of prior art, but provides the ability to conceive and test radical network designs and configurations. The actual simulation is by discrete processes or iterative replication of (randomized) events. Ten seconds of network simulation may represent several CPU hours.

One very important aspect of simulation to realize is that the device models are based either on deterministic or stochastic (that is, statistically random) processes. Deterministic processes represent sterile and primitive extrapolation, whereas stochastic modeling factors in the randomness and burstiness of network loads and interrelated processes and is thus much more desirable. In order to achieve an accurate representation of enterprise network performance, you probably need to create hours of simulated network performance so that the peculiarities of the random traffic generators and discontinuities of the mathematics do not skew results. Some models refer to this requirement as letting the model "settle down" or reaching an "equilibrium state." This is valuable advice as critical interactions and bottlenecks—particularly in a complex, multiple-path enterprise network—may be revealed only after many hours of simulated network activity have elapsed. This may correspond to a week or more of elapsed real-time. Figure 4.34 shows the startup spike resulting from a HyTech NetGuru simulation that has not reached equilibrium. When the throughput and utilization graphs show a repeating pattern—which they will—you will know that the model has reached the first useful points.

Results are presented in the form of bottlenecked paths and devices, bandwidth utilization, and response times. There are many discrete process network simulation modeling tools, some with GUI front-ends such as Comdisco BONeS, as shown in Figure 4.35. The network devices are mirrored by extrapolated objects that simulate the performance of the real thing. The simulation model performs well when the logical devices accurately represent the real ones, and when your model accurately reflects network loads.

The primary advantage of a state model is that the model reaches equilibrium very quickly based on user-input values. There are almost no calculations, and no statistical distributions of throughput processing based on traffic arrivals or process time distributions. State modeling is easily created with simple code; thus, bandwidth utilization results are almost instantaneous. State models by themselves cannot provide latency information, as that requires solution of complex simultaneous equations apart from the state model. Make Systems provides a wish-list of features in their six-part design, data capture, analysis, modeling, and planning toolkit; it is quite a complete environment, requiring a Sun SPARCstation. The WAN component also optimizes tarriffs for WAN services, a valuable optimization service for estimating the lowest transmission rates and limiting service requirements.

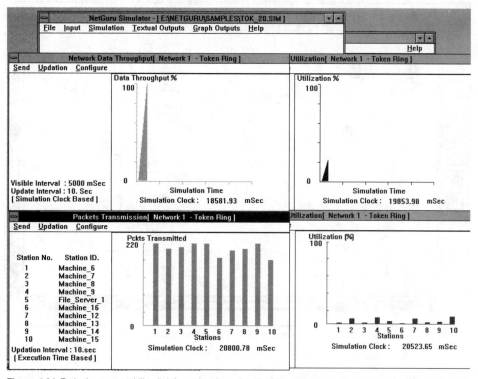

Figure 4.34 Early (not yet stabilized) information from the NetGuru simulation model of a complex Token-Ring internetwork. Nonetheless, the screen shots do show the type and format of the information you can expect from a good simulation model.

The primary limitation of state modeling is that all the conditions must be included by the designers to paint an accurate prediction of enterprise network performance and capacity. For example, there are link-failure states with routers that cause the hot standby unit to remap the network routing configuration. The router protocol (i.e., TCP/IP, IRGP, or OSPF), the network architecture, and the network load all determine how long it takes for the routers to rebuild the routing table and achieve router table convergence. Models of transmission links typically have two states, "available" or "not available;" but you can also add states for noise, intermittent service, long latencies, unidirectional service only (and which direction?), and different service levels and bandwidths. Every device on the enterprise network represents many more states than just "available" or "not available," and if these were included into a state model, you would have virtually infinite permutation. Infinite model space creates problems with model stability and increases calculation time beyond available resources. For obvious reasons, finite models simplify the real world.

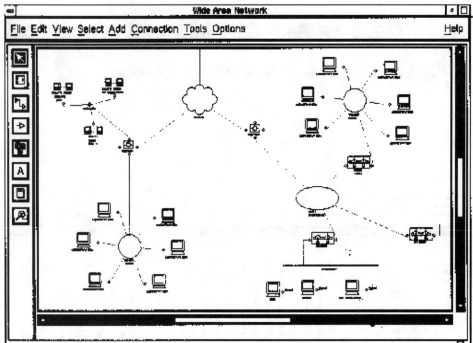

Figure 4.35 A screen shot of the graphical representation of a BONeS simulation model.

Statistical modeling

Statistical modeling is the process of defining traffic streams as mathematical distributions, defining network processes as other distributions, and then building simultaneous equations that the computer solves to reveal bandwidth and latencies. The equations are typically nonlinear (exponential) and quite complex, and mirror the arrival and process mechanics of network traffic dynamics very well. Fractal mathematics and self-similarity, as mentioned in Chapter 2, look even more promising than simplex theory, operational research, or queueing theory as a basis for modeling network performance. However, I do not know of any performance modeling tools based on fractal theory yet. Most analytical tools use simplex simultaneous equations, and the more significant ones apply queueing theory because they create solutions in practical time on average computers with statistical accuracy.

Statistical modeling generates the same results as simulation. However, with statistical modeling, you can additionally solve the equations to determine minimal resource requirements, performance breakpoints, or optimal performance configurations. I like it better for those reasons. I also like it better because results are immediate, and a good model does not need to stabilize over hours or days. You get more information out of the model faster (and often with higher reliability) than with simulations, because you are less dependent on the accuracy, stability, and behavior of the simulation objects. By the way, Make

Systems uses queueing equations to solve for transmission latency, which the state-change model cannot provide.

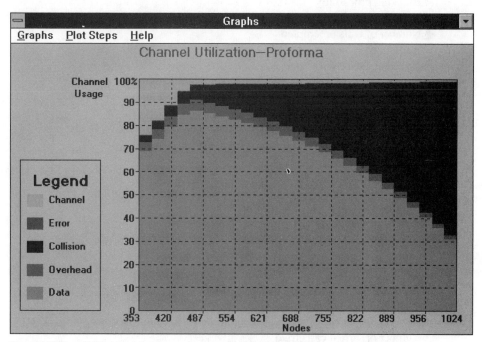

Figure 4.36 The LANModel graph shows performance breakpoints for different configurations.

However, statistical modeling usually comes with a higher cost—in terms of overall complexity, confusion, and difficulty in specifying a model representative of the enterprise network. Typically, it is quite difficult to gather the information necessary to define the traffic arrival distributions, the device processing times, and the mechanisms for the buffers (thus queues) and overflows. On the other hand, queueing models can be designed so that they are not so sensitive to bad or insufficient data. You might refer to *LAN Performance Optimization*, which has a disk containing the LANModel traffic queueing model. Figure 4.36 shows a typical bandwidth display with obvious performance breakpoints.

To effectively design, build, gather data for, and interpret an enterprise network model correctly, you should understand (or at the very least be familiar with) the following statistics:

- Percent of circuit availability
- Number of circuit group failures
- Number of busy circuits (average, standard deviation, maximum)
- Bandwidth used
- Utilization percentage
- Number of calls attempted
- Number of blocked calls
- Number of calls carried for each end-to-end path

- Number of calls blocked by link group failure
- Number of calls blocked by link group traffic
- Blocking probability
- Number of calls carried and disconnected
- Number of calls preempted
- Number of calls queued
- Queueing probability
- Call queue size (average, standard deviation, maximum)
- Call queue time (average, standard deviation, maximum)
- Number of packets transmitted
- Buffer use (average, standard deviation, maximum)
- Packet queue time (average, standard deviation, maximum)
- Number of packets processed
- Number of packets blocked
- Packet switch wait time (average, standard deviation, maximum)
- Number of average busy processors
- Switch utilization percentage
- Number of messages blocked by the node input buffer control
- Number of whole messages sent and received
- Message delay (average, standard deviation, maximum)
- Total number of packets delivered
- Average packet delay
- Network throughput (aggregate over all message categories)
- Number of virtual calls tried
- Number of virtual calls blocked
- Number of virtual calls rerouted
- Virtual call setup and access delay
- Number of calls ended
- Average call length
- Availability percentage
- Number of collision episodes
- Number of collided packets
- Number of CSMA/CD deferrals
- Deferral delay (average, standard deviation, maximum)
- Deferral queue size
- Number of multiple collision episodes
- The maximum number of collisions per episode
- Number of packets delivered

These quantities are usually resulting outputs. To understand a simulation with such results, you really need to understand the numbers and their meaning. More than likely, you will need to define your devices and subnets using such terms and values as well, so as to build a usable model. There are alternatives to learning how to use these tools yourself, as vendors typically create a front-end to mask the complexities of the simulation models and provide output already analyzed by software. LANModel is specific to LAN and inter-LAN modeling, but provides some analysis of an Ethernet segment with 97 percent traffic loading and latency over one second, as Figure 4.37 illustrates.

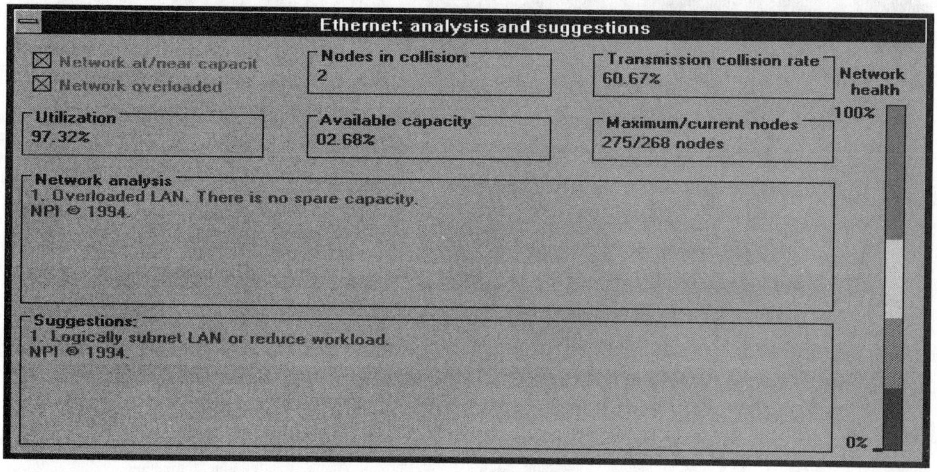

Figure 4.37 A screen shot from the LANModel analysis and suggestion report.

Another approach is put forward by Advanced Visual Data, a company that will engineer a simulation model custom-designed to your telecommunications and data communications environment. Figure 4.38 shows a screen from their Advanced Visual Data simulation model designed for a trade show presentation.

With much the same method, IBM Network Services is marketing network design and optimization services based on research originally performed in Yorktown, NY. The Intrepid modeling system (based on queueing models) has now become QANDA, a service that uses the Intrepid tool to design and analyze router, N-way networks, and IDNX switching backbones. The results are based on steady-state modeling—so that it cannot model router or link failures—but it can dump its information into Snapshot, a discrete simulation tool, that will extract the data throughput from the overhead component in router traffic. One of the more interesting aspects of QANDA is that it can provide optimization of tariff rates for domestic and international tie-lines and WAN services, using standard service provider databases.

Emulation modeling

Emulation is another type of modeling that I almost forgot to include in this book. It is very valid and useful for a preexisting enterprise network. Emulation is the process of imitating (simulating) an environment to see how a real one might perform. This means you really build the network with physical components and infrastructure that mirrors what you hope to model.

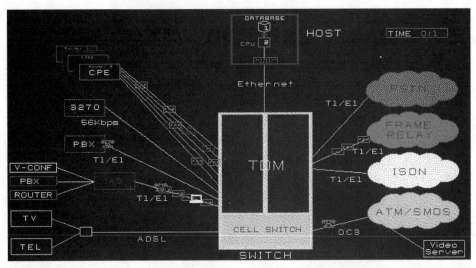

Figure 4.38 A screen shot from a custom-designed Advanced Visual Data simulation model.

There is more to emulation than just the physical infrastructure and network nodes. You also have to add software, traffic, and processes. Emulation is typically performed in laboratories (as with Ziff Labs or InterLabs), vendor research and design facilities, as part of a request for proposal (when it is a really big network), or at trade shows. Most other organizations do not have the staff, expertise, resources, or time to create emulations that reflect a true environment.

On the other hand, there is another type of emulation that many organizations can do and often try. This really is a physical extrapolation of the current environment. Refer back to the extrapolation example with 60 users at 60 percent capacity. The organization wants to know how many more users it can add doing the same workload. If you have similar hardware in use for other operations—instead of the order entry, perhaps it is finance or legal department equipment—you can temporarily move it, install it, and add the extra load during a weekend or nighttime test. You will know when you reach capacity.

I have one serious warning about the efficacy and accuracy of emulation modeling. When emulation uses test scripts, captured keystrokes, or canned captured packets, you may not be creating a true emulation environment. Scripts and captured keystrokes represent prior work that, when duplicated, may already exist and thus may not be performed with the same loads and overhead as with the original process. Additionally, when capture packets are inserted or injected (with a protocol analyzer) back onto the network, this load originates from one point, not the multiple points that it originally came from. This creates two modeling errors. First, the single-source injector is not going to fight for bandwidth, experience collisions, or experience latencies comparable to the multipoint source. Second, the packets are duplicates of prior messages and workloads that may not be appropriately related to any current processes, applications, or loads, and so may merely be dropped by bridges, routers, and gateways, or simply ignored by servers and workstations.

Capacity planning

Typically, LAN managers are unfamiliar with capacity planning, unlike their counterparts running the "big iron" mainframes. Historically, the costs for LAN equipment were not so great and experience with LANs not so lengthy that capacity planning became a critical activity. Also, each succeeding generation of hardware provided a 10-fold increase in capacity, so as to provide inexpensive solutions for performance bottlenecks. Even now, the costs for equipment and infrastructure represent 15 percent or less of the fully burdened costs for enterprise networking. Now, however, each new generation seems to provide only about 1.4- to 2-fold improvement over the prior ones, and most of the capacity bottlenecks are not even remotely addressed by faster hardware.

As LANs now support mission-critical tasks, replace mainframes, and become enterprise bottlenecks, capacity planning is a critical skill. This is an important—or even essential—skill for designing, managing, and tuning the performance of the enterprise network. Network capacity planning is a specialized aspect of network modeling. Although you typically employ extrapolation, simulation, and statistical modeling to calculate network capacity, it is important to recall the concept of the critical path from Chapter 2. Capacity planning is the modeling of critical paths, those paths without slack.

The critical paths that constrain capacity are not necessarily the paths with the slowest throughout. If you look ahead two sections (see "Switching vs. routing" on page 160) to the comparison of the performance of bridges and routers, an ancillary result of the model is that enterprise network is that the FEP is probably the first device to become a capacity limitation. You need to design a model that stresses your critical paths and exercises a real test of potential bottlenecks. As another example, simply to conclude from a bandwidth model that FDDI or Fast Ethernet will outperform the currently bottlenecked 802.3 Ethernet infrastructure does not account for the effects of latency—the latency differences for signal transmission over the wire or fiber between Ethernet and FDDI are almost nonexistent—or represent the ability of network devices to generate more than 4 Mbits/s of real workload to require the greater bandwidth. Critical path capacity modeling requires a very detailed and carefully constructed model.

Bridging vs. routing

Bridges are simpler and faster. Routers are slower, more complex, support multiple protocols, and create firewalls. Which one is better? The decision is not always a function of speed; there are software issues too. Sometimes the time required to manage a large, routed internetwork is more than an organization can withstand. In some situations you cannot route traffic; for example, NetBEUI is not routable. Also, bridges support SLIP or PPP on-demand dial-up connections (to multiple sites), which routers cannot. How do you model the performance of a bridge that provides compression? How do you quantify the effects of filtering broadcast traffic, as is necessary for high-level protocols? Sometimes you cannot justify the added purchase price for routers. It is not always clear either that bridges create panics and extra traffic, or need to be replaced with routers. The vendors cannot guarantee which will work better. Although one might say that bridges are better for remote WAN connections and routers better for enterprise networks, such blanket statements are false.

Instead, let's model this process. There is a packet source, a queue, a processor, and completed task. Queue size and dropped packets are a function of packet sizes and device buffer sizes. Processing time for each device is primarily a functional of packet size. The model may look like the queue in Figure 4.39.

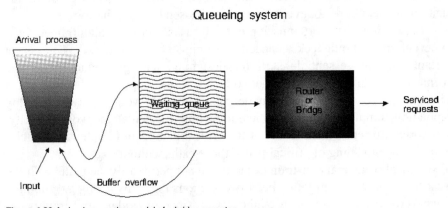

Figure 4.39 A simple queueing model of a bridge or router.

There are extra complications. First, both bridges and routers are bidirectional servers with arrival distributions from two networks. Secondly, it is not really a single-stage process, because task completion is not just a function of the speed of the bridge or router, but also of the state of the recipient network. If the bridge or router cannot get the token (FDDI and Token-Ring) or transmit without a collision (Ethernet), the task fails. So, in effect, each bridge or router represents four stacked queues (or more if they support multiple ports or protocols), as Figure 4.40 indicates.

When you do not model the environment and process correctly by simplifying the bridging or/and routing process, the bridge with a faster process time will always outperform the router in your model. Effectively, the router is providing the same service, but with a slower service time. This isn't what you want, nor is it very predictive for your future. You can simplify the enterprise network under some conditions, but be careful when oversimplifying the intermediate nodes and application messaging processes.

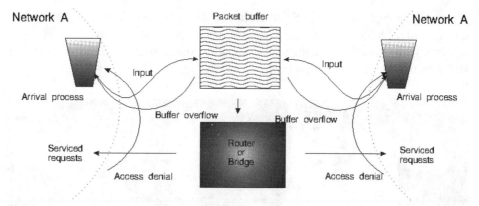

Figure 4.40 A more realistic queueing model of a bridge or router.

There is one more complication with the router. Not all the packets are forwarded; some are filtered or dropped under excessive loading. This is represented as a ratio on the input and rescaling the arrival distribution, or simply as a lessened input stream. The following screen shot from Prophesy! in Figure 4.41 shows a queueing simulation result for just one level of network loading and packet filtering. In fact, this model does simplify the structure of the enterprise network, but not the bridge and router.

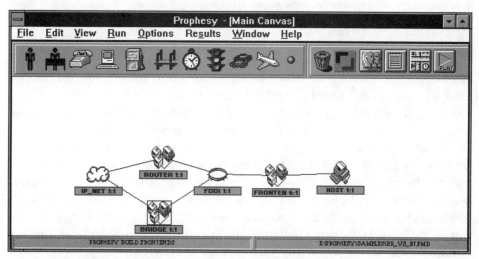

Figure 4.41 A Prophesy! model of a bridge versus a router on a simplified enterprise.

The Prophesy! simulation yields the parameters for the router and the bridge, as shown in Figure 4.42. Notice that for the same workload, the router exhibited saturation, while the bridge showed a maximum utilization of 25 percent. Clearly, the router could not keep up with the traffic going over the bridge. This does not mean that the router is not good and could not do the job intended, that the bridge is better, or that the network has a bottleneck. Understand the limitations of modeling and your model in particular; this model does not answer the question of which of the two devices is better, except for the environment represented by the model's scenario. In fact, the router performed as a bridge or simple repeater and filtered no traffic.

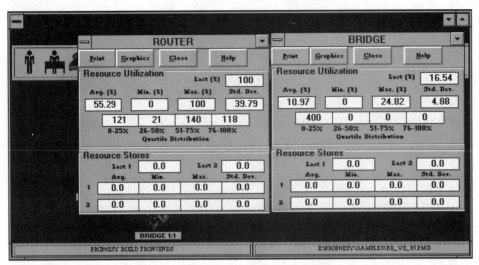

Figure 4.42 Prophesy! device load information showing that the bridge is a better performer when paired with a router that filters no traffic. However, you might notice that this complex network—the complexity is actually hidden with device counts behind each object—is surprisingly more likely overloaded by the front-end processor than it is by the bridging and routing connections.

I used this model for a purpose; that is, to show the benefits and limitations of any model. If you wanted to build a better model strictly to compare and contrast the bridge versus the router, you would modify the dual-path model with one active single path through the network; *either* your bridge *or* your router, but not both. Do two simulations until the results stabilize and then analyze the statistics. Results are valid only for the static comparison between the two devices for that network design and modeled traffic levels. Nonetheless, this single point-to-point path represented in this modified model is not representative of a true enterprise network, which would have more than a single route active, as is true with the original. The queue utilization numbers in Figure 4.43 confirm the suspicions that the bridge is faster than the router, but basically show nothing new. The fact that the queue values are mostly zero perhaps suggest that the network model was not sufficiently detailed, accurate, and complex enough, or perhaps not even designed to correctly reflect the environment I wanted to model.

The lack of exact fit of the model to your environment is a common shortcoming with general-purpose queueing tools. Prophesy! is actually quite good for modeling generalized problems such as traffic, business workloads, and customer service utilization. There is significant statistical support behind the simple front-end, but as a result (or in spite) of that front-end, you really do need to understand statistics, modeling, queueing theory, and designing simulations. For those who know, it is a good general-purpose queueing model.

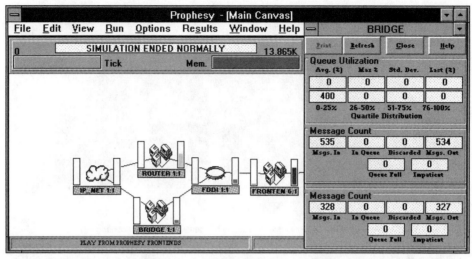

Figure 4.43 Prophesy! queueing information. The bridge received almost twice as many messages (535) as the router (328), confirming that the router was overloaded and spilled load to the parallel bridge processor.

On the other hand, the enterprise network is a specific and complex traffic queueing problem. Prophesy! will model the enterprise if you build a good model and collect sufficient data to define the many variables. You will discover that it is very hard to match the model parameters to the actual environment you wish to model. The various queueing parameters, arrival distributions (Poisson, normal, hyper-normal, or Erlang), queue lengths, and methods for handling buffer overflows and network failures, have a profound effect on the model results. Slight changes yield significant differences, and you cannot rely on the default values or assumptions (from the tool designer) because they will not be representative of your network environment. You will need to review and justify every parameter before you see any results. You want to create the model with as little as possible of the bias that might occur once you have seen the results.

I am not yet finished with the bridge-versus-router problem. This problem can be solved with Prophesy! or with GMS 2.0 (Probots). GMS, which is DOS-based, was faster for performing the multiple matrix calculations necessary for a more global solution than the single point solutions from Prophesy!. I also adjusted the network modeling parameters so that the bridge and router overflowed, rebuilt routing tables, and stuttered with destination backups. Realize that you have two different arrival streams with different distributions going into two other mergent streams with two other different distributions and various forwarding and filtering rates. These can be plotted on the same graph to yield the following performance curves for bridges versus routers (with selective packet filtering by destination address enabled). Figure 4.44 shows how performance varies with an average packet size of 230 bytes (with a normal distribution) and with the router forwarding 20 percent of all packets received.

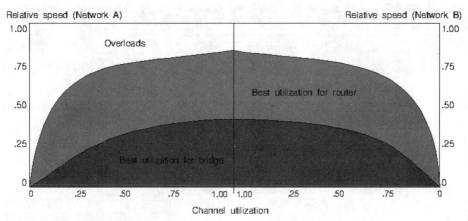

Figure 4.44 A performance plot (with GMS results) for a single bridge versus a router scenario.

The actual breakpoints for selecting a router instead of a bridge depend on base network traffic loads on each end of the router, the actual packet forward percentages (that is, how much traffic is filtered and dropped, and how much is routed onward), the traffic levels directed to the router and the arrival time distribution, the router processing speed, network traffic speeds, router buffer sizes, and the packet size distribution. If you still believe that the decision between a bridge and a router can be trivialized, refer to Figure 4.45 to locate the scenario in the previous figure in the full parameter space.

Switching vs. routing

Switches are faster, right? It all depends. Routers are slower, more complex, support multiple protocols, and create firewalls. Which one is better? The decision is not always a function of speed; there are software issues too. I decided to model this comparison, and had no preconceptions as to the results. The switch replaced the bridge in the prior Prophesy! model. Its parameters include 50 ms switching time, four-fold faster than the router and twice as fast as the original bridge. However, the switch was defined so that traffic overflows were not buffered, and packets were not filtered at either the MAC level or network layer. The results were too sensitive to the input parameters, and the model was not sufficiently stable to make any useful assessments. When it worked, the switch was four times faster than the router and twice as fast as the bridge; otherwise, it did not process any packets at all. The problem wasn't the model, but rather the parameter space.

Using GMS, the switch replaced the bridge in the bridge vs. router model. Switch definitions were set as described on page 160. I also adjusted the network modeling parameters so that the bridge and router overflowed, rebuilt routing tables, and stuttered with destination backups. You have two different arrival streams with different distributions going into two other mergent streams with two other different distributions, and various forwarding and filtering rates. Figure 4.46 shows how performance varies with an average packet size of 230 bytes (with a normal distribution) and with the router forwarding 20 percent of all packets received. The switch forwarded 100 percent of packets directed to it.

The decision is not so obvious. The switch outperformed the router at lower traffic loads. Throughput dropped with increased traffic loads and the router outperformed the switch. The ability to buffer packets and match the forwarding rate to the destination network is clearly an important factor. The surprise might be that as traffic loads increased, switch throughput dropped. When the switch was flooded and blocked from forwarding packets, it suppressed the network. Source quench (where lack of message acknowledgment provides an inference that the network has a problem and that the source nodes should stop transmitting) might level the switch curve performance. But this does not really address the condition where the slower router outperforms the switch. The conclusion is that router (or store-and-forward switch) buffer size has a bearing on enterprise network performance.

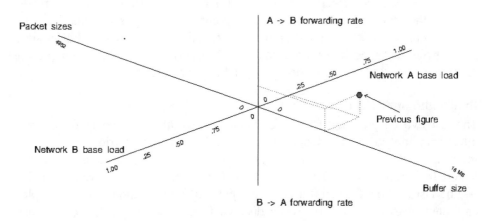

Figure 4.45 A simplified parameter space for a router modeling configuration. The dot shows the parameter choices used in the prior figure.

Router buffer size

If the switch performed badly because its buffer size was 0 MB and the router had a store-and-forward buffer, the question is what is the optimum buffer size for a router? Although GMS really requires a good handle on mathematics to use it, it seemed a good bet that the tool might answer the buffer size question.

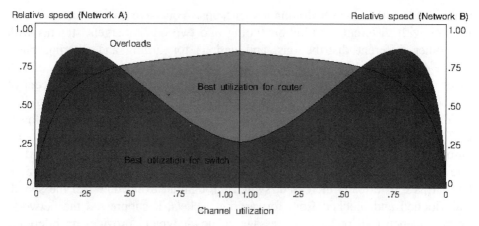

Figure 4.46 A performance plot (with GMS results) for a single switch versus a router scenario.

Unfortunately, the parameter space—packet arrivals, processing time, packet size distribution, and network loads on both Network A and Network B—provided too much instability and correlation. The following graph shows a result by assigning an average packet size of 200 bytes and steady-state network traffic loads. Clearly the parameters matter, and the relationships among them do too. The result is not specific enough to be particularly useful, but does fit the concept that buffer space matters to a point. The only scale that actually generated useful results was buffer size as a function of packet size; buffer sizes in kilobytes and megabytes were unstable. Not surprisingly, the graph (Figure 4.47) looks much like a performance vs. cache size measurement.

Bandwidth and latency

The last two chapters tried to differentiate bandwidth and latency from many other commonly used LAN performance statistics, and also tried to stress the importance of latency as one of the most critical factors for enterprise network health. In fact, these two traffic parameters are the ones you really want to model. As the last section showed, they can be quite difficult to model for a complex environment. Any of the tools so far discussed can model one LAN, one aspect, one bottleneck, or contrast the performance of comparable devices. It is more difficult to handle the nonlinear aspects of bandwidth utilization and latency enterprise-wide. You need more sophisticated models and tools.

So far, the models were built with traffic information captured at the physical layer. This is represented by the packets, bits, bytes, and transmission frequencies. This may include the MAC- or protocol-level of networking. It is actually more reasonable and accurate to model enterprise network performance at a message- or process-level. This corresponds to a data-, network-, or application-level load measurement. Figure 4.48 shows a COMNET message latency display.

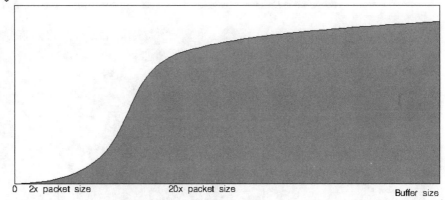

Figure 4.47 Router throughput as a function of packet buffer size.

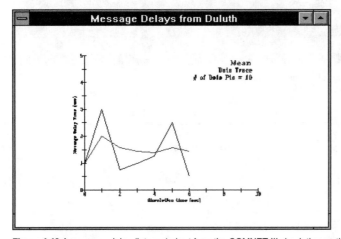

Figure 4.48 A message delay (latency) chart from the COMNET III simulation on the CD-ROM.

COMNET uses traffic generators based on random arrivals, triggered messages, or user-supplied data (captured data or trend analysis). Parameters include class of service, priority, type of service, and how traffic overflows are handled. Figure 4.49 shows a sample enterprise network drawn in COMNET III.

The two graphs in Figure 4.50 show the latency and bandwidth utilization for 512-byte messages in Ethernet and Token-Ring. BONeS can also model message streams in addition to lower-level traffic. I placed them side-by-side with the same scale, so that you can compare the performance characteristics of these two common LAN protocols. What you need to notice is that message size also varies by application, network environment, protocol, and network operating system. The calculations that form the basis for these are very specific in that the message size was fixed, rather than allowed to vary within a range.

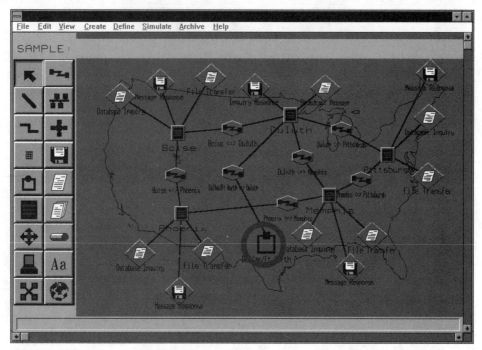

Figure 4.49 A COMNET III simulation of an enterprise network.

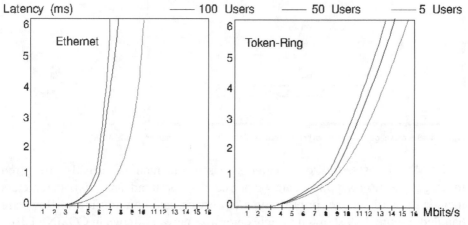

Figure 4.50 Latency vs. bandwidth for Ethernet and Token-Ring generated with BONeS PlanNet simulator (Network Performance Corporation, Dunkirk, MD).

Reliability and survivability

Predicting network reliability is actually a very difficult statistical task. Network reliability consists of the combination of all in-line components between source and destinations. In-line components usually consist of all interconnection devices, servers, hosts, and output devices. The prediction is usually couched in terms of mean time between failure (MTBF), which is a predicted average value with a variance about the mean. These numbers are raw values.

A router with a MTBF of 8.4 years does not mean that a router will work without flaw that long, but rather that the hardware is unlikely to suffer a mechanical failure in that time. A routing table might have a MTBF of only 8 days, while the software or firmware may not be rated at all. Simply to assign the lowest MTBF value to the network on the assumption the network is only as strong as the weakest link is erroneous. While that calculation may seem intuitively right for a LAN (it isn't), it certainly is statistically unsound for an enterprise network with multiple paths. Even a RAID unit with two disks, each with a MTBF of 8 years (or failure rate of 0.003 percent), would experience a rate of failure of a single disk of the same 0.003 percent. You would expect a failure rate of 0.0000001 percent for the disk drives as a paired unit, as shown by the serial failure rate equation below:

$$F_1 = \text{Rate of failure for path 1}$$
$$F_2 = \text{Rate of failure for path 2}$$
$$F_{1+2} = \text{Rate of failure for both paths serially}$$
$$= F_1 F_2$$

You may find it hard to calculate the failure rate for a key disk sector on one of the drivers because the input is hard to find. It follows that it is difficult to calculate routing table application software failure rates. Additionally, the effect of a failure on one path only means that the secondary or standby routes are activated, and are used to bypass the failure. Calculating the failure rate is the chance that both routes fail simultaneously, which is not the sum of the two rates or the lowest rate, but rather a complex combination:

$$F_1 = \text{Rate of failure for path 1}$$
$$F_2 = \text{Rate of failure for path 2}$$
$$F_{1\&2} = \text{Rate of failure for both paths simultaneously}$$
$$= 1 - ((1 - F_1)(1 - F_2))$$

It is possible to calculate critical failure rates and spare equipment inventory or hot standby requirements. In fact, LANBuild (LAN Designer) creates a report for networking component inventory requirements based on statistical (historical) failure rates and device counts for different types of protocols and computational environments. See "Spare parts" on page 111 for such a report.

Predicting survivability is actually a very difficult statistical task because it entails not only forecasting the failure rates, but also predicting how the enterprise network and the organization ultimately reacts to the failure. This is certainly the more interesting and relevant prediction of the two. This is what managers really need. What it entails is constructing a list of all possible events, calculating the odds that they will occur, and then constructing a plan for coping with each event. (See "Disaster recovery planning" on page 112 for additional information.) For example, a critical router can be backed up by the standby router; usually, however, it is a little more complex than that. You want to model the effects of component failure and see how the infrastructure responds.

Model network performance and inject events that may incapacitate the network, and explore the reaction. Prophesy! allows you to disable any device in a model by clicking on it with the right mouse button. This can be useful in some instances, as Figure 4.51 shows.

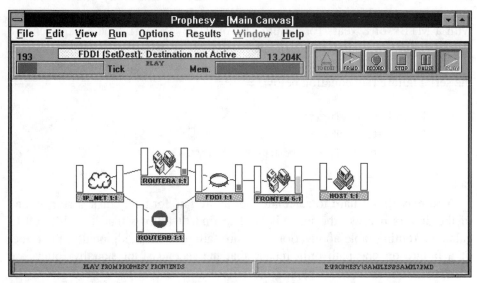

Figure 4.51 The effects of a router failure show the endemic traffic throughput reduction on the enterprise.

However, Prophesy! cannot support the complexity of interaction on an enterprise network, when devices fail, without user programming each result. It is difficult to model the cascade and subsequent network panic without defining and creating event-driven processes triggered by objects, traffic levels, alarms, or other events, but it is sometimes possible, as Figure 4.52 illustrates.

The more expensive and expansive tools do not necessarily have the logic in place to handle the effects of a device failure; you may have to manually create traffic tables to replace the standard traffic generators, and define events that occur under specific conditions. When a gateway fails, as in the prior figure, reconfiguration or activation of a hot standby router in parallel to the failed unit may require a minimum of five minutes; it takes more time on complex internets. Although Prophesy! could inject a five-minute delay in this situation, when a router fails and more indirect routes are required, the reconfiguration and network effects of rechanneled traffic are actually more complex than just a five minute hiatus. There is that cascade of effects as sessions drop, processes attempt retransmission, and users give up and go home. Consider also the case of a failed server. Although a standby unit could automatically kick in and update the state of the RAID disks, what happens when NetWare fails to understand the reconfiguration, and does not let users log in because the password services have failed? You can model the cascading effect of certain device failures only if you can codify and quantify the effects in such a manner that they can be included as part of the model.

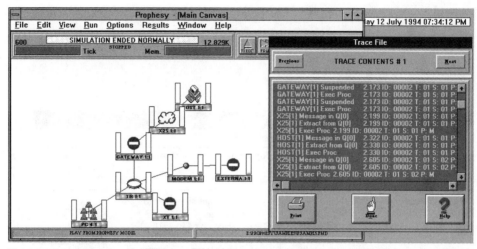

Figure 4.52 An initial failure at the gateway, which is in-line to all network traffic, represents a critical failure with profound secondary effects as the computers on the Token-Ring and the remote PC also fail.

Simulation of packet size

Frame and packet sizes, as well as cell sizes, have a profound effect on network latency and bandwith utilization. As Chapter 2 asserted, larger packets are more efficient, but are more apt to create buffer overruns and bottlenecks at intermediate nodes. This section substantiates this assertion with a LANModel statistical model provided with the *LAN Performance Optimization* book.

Specifically, when bandwidth utilization levels are the same for two samples, the sample with the larger packets creates more burst and peak overload conditions. By the way, you can also model the case that shows that you achieve higher channel utilization with larger packets; this merely confirms the previously-referenced router test results from Interlab and others. It is neither interesting nor pertinent. The goal is not to pump up the bandwidth utilization on the enterprise network, as that represents empty primping and preening; but the goal is, rather, to achieve increased work accomplishment. Figure 4.53 illustrates bandwidth utilization at 13.3 percent with packet sizes averaging 118 bytes, a reasonable level for an Ethernet segment, while Figure 4.54 illustrates the channel bandwidth utilization growth potential with that same packet size.

You can improve network throughput five-fold growth to around 5000 packets/s, as this graph illustrates. Latency increases disproportionately, with profound segment traffic jams and backups to interconnected segments. The figures later in this section show the increased latency values. On the other hand, Figure 4.55 also shows bandwidth utilization at 13.3 percent with packet sizes averaging 572 bytes, still a reasonable level for an Ethernet segment, while Figure 4.56 illustrates the channel bandwidth utilization growth potential with that same packet size. Notice that sustainable network traffic loads peak around 2000 packets/s. The curve for the latency goes ballistic around 2000 packets/s per second, which you can corroborate by playing around with LANModel.

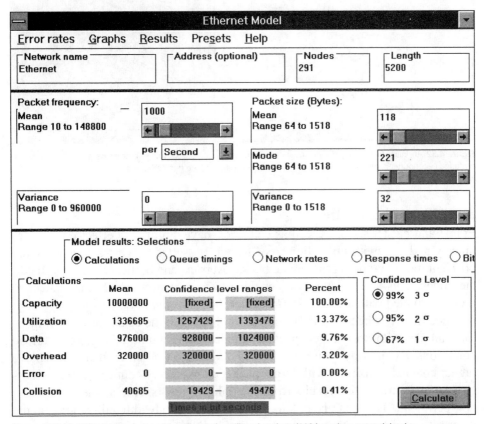

Figure 4.53 A LANModel simulation of the effects of small packet sizes (118 bytes) on network load.

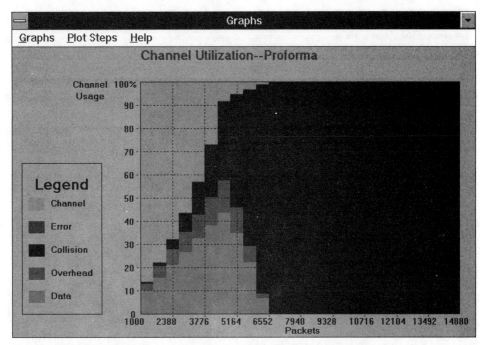

Figure 4.54 A LANModel proforma graph of increased network loading for small packet sizes.

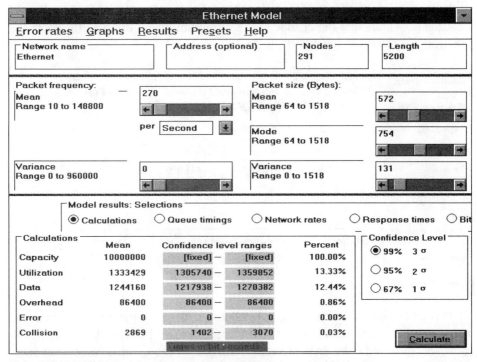

Figure 4.55 A LANModel simulation of large packet sizes (572 bytes) on network load.

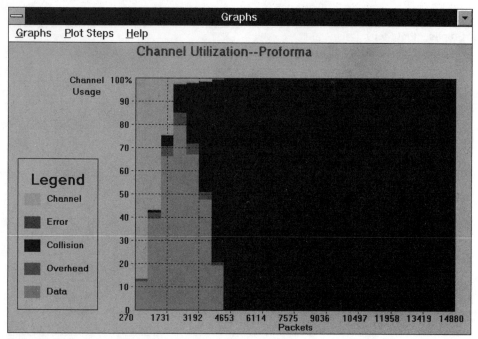

Figure 4.56 A LANModel proforma graph of increased network loading for large packet sizes.

The following two figures, Figure 4.57 and Figure 4.58, illustrate the utility of the latency. Notice that the change in the minimum queue timings illustrates the loss from the overloaded network channel.

Queue timings	Mean	Confidence level ranges		Percent
Minimum	1120.0	1072.0 —	1168.0	
Expected	19.5	17.5 —	21.5	
Deferral	1120.0	1072.0 —	1169.2	
Total	1139.5	1089.5 —	1189.5	

Figure 4.57 Queue timings from LANModel.

Queue timings	Mean	Confidence level ranges		Percent
Minimum	12320.0	11991.4 —	12320.0	

Figure 4.58 Queue timings from LANModel at a far higher bandwidth utilization.

This particular queueing analysis has additional utility for capacity planning. You know that transaction growth will increase traffic bandwidth requirements. However, the bandwidth growth itself consists of some distribution packet sizes and arrival times, and if you ignore this, you are apt to overload the enterprise. As you can see, you will need to model not just the bandwidth requirements, but also the two components that define the traffic bandwidth distribution. Failure to adequately forecast capacity in this way could put your enterprise at full occupancy even the average bandwidth utilization seems insignicant. This shows the statistical divergence of the traffic averages from the effects of the variances.

LANModel transaction modeling

Although *LAN Performance Optimization* included LANModel 1.0, this software was primarily a MAC-level predictor. Additions to LANModel 2.0 include the transaction load and enterprise modeling modules. The transaction modeling component converts an on-line transaction processing load (TPS) within an SNA environment into equivalent LAN protocol message loads, and then converts this to MAC-level loads, as Figure 4.59 shows. It calculates straight loads, loading with RAID or write-behind disk arrays, and effects from rollbacked databases.

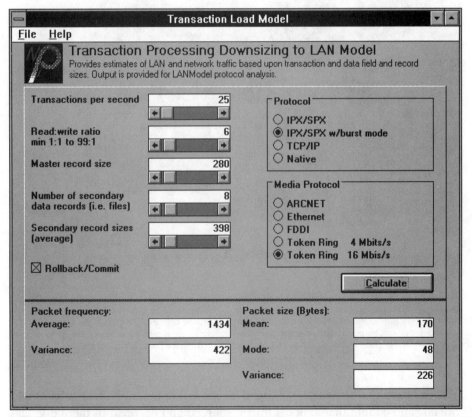

Figure 4.59 The LANModel TPS module converts a standard TPS measurement into a statistical MAC-level load for a LAN-protocol-driven network. The results are in terms of average and variance for the queueing model.

Enterprise network cross-product traffic

The problem of modeling the enterprise network is not that it is composed of subnets and segments, but rather that these subcomponents generate traffic locally and then export it to some or all of the other segments. This traffic between them, a cross-product, is the difficult factor. It is quite easy to model the gateways between subnets, segments, and hosts—it is just a standard bi-directional queue—and string the subnets and segment models together into an enterprise model. Why enterprise models often fail is because the designer fails to account for the cross-product traffic. Nonetheless, logging the intersegment traffic represents the more difficult task; once accomplished, it really is an easy matter to add it into the enterprise network model. Figure 4.60 illustrates the atomic subnet and segment units, and the cross-product traffic.

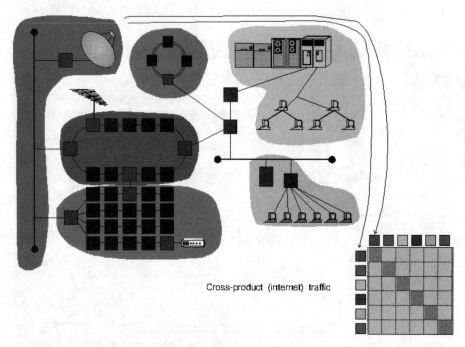

Cross-product (internet) traffic

Figure 4.60 The atomic units and internet traffic in the typical enterprise network. The chart in the lower right-hand corner shows the cross-product data collection necessary to model an enterprise network. The gray diagonal squares correspond to LAN simulation, whereas the other areas represent the possible internet traffic.

The solution to this problem is found by modeling the individual subnets and segments, and adding the nonlocal traffic sourced and destined to the individual subnets and segments; that is the cross-product traffic. Either tally the rows or tally columns in that chart, and use that load information for modeling the enterprise network. Alternatively, you can model each segment or subnet separately, but include the cross-product traffic to determine the real loads on them when they are attached to the enterprise network.

Conclusion

This chapter explored the types of tools required to capture the data necessary for analyzing enterprise network performance, modeling network performance, and tuning specific aspects of the enterprise network. This chapter has covered blueprinting, inventory tools, tuning tools, and both simulation and emulation modeling tools. You have learned how to use general-purpose tools for the specific task of performance optimization. Chapter 5 applies these techniques for performance optimization, and teaches the skills and technologies that are pertinent for tuning enterprise network performance.

Chapter

5

Optimizing
Networks

Introduction

This chapter presents techniques of enterprise network performance optimization as well as some specific methods for relieving the performance bottlenecks. The key section in this chapter also contains a series of simplified logical network designs showing progressive redesign and reconfiguration techniques for improving the performance of the enterprise network and reducing the effects of traffic bottlenecks; see "Optimizing with design" on page 179. Because bottlenecks represent complex traffic flow problems, this chapter focuses strictly on issues of design, bandwidth, latency, configuration, integration, and software.

The major concept for this chapter is that the enterprise network does not usually have singular active constraints which can be resolved with single-point solutions. Rarely can performance be affected with a faster protocol, better cabling, or improved applications, as might be the case with a LAN bottleneck. The enterprise is an infrastructure of related, distributed, and integrated activities, and typically benefits from multiple tuning efforts and fundamental design improvements. You cannot rightly expect to replace a single component and correct all major deficiencies in a single step. Enterprise tuning is iterative, and it is also a "big picture" activity. The key topics presented in this chapter are:

- Infrastructure fortification
- Optimization methods
- Optimizing with design
- Generalized optimization techniques
- Financial optimization
- Technical solutions
- Energy usage
- Optimizing desktops
- Optimizing work groups
- Optimizing LANs
- Optimizing WANs
- Optimizing network operating systems
- Tracking the disconnected device
- Optimizing application code
- Encoding applications to support WAN glitches
- Optimizing transaction processing
- Client/server transaction processing
- Servers
- Clients
- X Windows
- Anomalies
- Optimizing WAN connections
- Optimizing hub and spoke networks

Infrastructure fortification

The enterprise network is primarily a multistar technology with routers connected by T-1 lines when remote sites are part of the enterprise. The bandwidth of T-1 is limited in capacity for LAN-type traffic, and the 1000-router network is not necessarily the solution for providing the increased bandwidth and improved latency necessary for imaging, client/server, or multimedia. You need to fortify unique network components. One approach is to use alarms and a spare parts inventory. This is the cheap approach. Another approach is to build in redundancy. This adds complexity, but potentially minimizes the window for failure. The third approach is to migrate toward structured wiring, address the limitations of the infrastructure, and work within the infrastructure's inherent capacity constraints. In many cases, there is no way to sidestep inherent limitations.

Structured networking is vendor-created catch-all phrase used mostly to encourage you to design networks with parts from particular vendors. It does not mean much; perhaps it suggests the concepts of plug-and-play networking. While that is certainly a desirable concept, it is far from reality today. However, *structured wiring* implies adherence to the concept of installing multipurpose copper (and fiber) plant, according to a structured plan for telecommunications and data communications. Structured wiring provides interchangeability and options for the enterprise network.

Although structured wiring initially may seem more expensive, the flexibility, increased reliability, lowered management and maintenance costs, and the ease it provides to patch around problems is very rewarding. Antique cables and retrofits often create performance, configuration, and integration nightmares. If you are considering upgrading the wiring plant incorporate the concepts of EIA/TIA 568/569. It provides flexibility and redundancy, which for the most part are essential for mission-critical activities. Also, structured wiring usually implies that someone has created a systematic and detailed wiring diagram—a diagram that will be critical to optimizing enterprise network performance. Such schematics (as shown in Chapter 4) provide the basis for qualifying all the network segments, for providing options and work-arounds, and giving many degrees of freedom for testing different configurations. Without such schematics, you cannot take a broad view of the network and optimize routing. This represents the fundamental approach for optimizing enterprise network infrastructure bottlenecks.

Optimization methods

The process of optimizing the enterprise network—whether it is a collection of interconnected LANs, LANs and WANs, hosts with gateways communicating with servers and clients, or some other mix of structures—depends significantly on configuration, purpose, and goal. You also want to resolve performance bottlenecks before they escalate and make the enterprise less productive and more trouble-prone. This is not only a technical issue, but also a public relations concern as well. The perception of network bottlenecks to users and key personnel is as important as the reality of performance obstacles and tuning problems. Here is a word of caution: Just as with LANs and computer systems, on the enterprise network you optimize what you seek to optimize. This is true whether or not you have picked the correct goal, interpreted the benchmarks results correctly, or identified the true sources of performance bottlenecks.

For example, if you assume that bandwidth is the bottleneck because of all the protocol analyzers show nearly 100 percent channel utilization, and therefore replace 10Base-T with FDDI, the result may disappoint you. If you optimize a process path that is not the bottleneck, you will see no performance improvement. If you optimize a process path that is an active bottleneck and expect a 10 percent speed gain because you increase that component's speed by 10 percent, consider these two qualifications. First, adding slack to the current bottleneck may only cause another component to become the active bottleneck. Second, improving performance of a single component by 10 percent does not always provide a direct correlation with computational results.

It is true that this component will perform a task 10 percent faster. Because this component represents one path of many and is only a linking path, the project is unlikely to be 10 percent faster. More likely, while this previously critical path may possibly operate 10 percent faster, some of the 10 percent path savings may become slack time. In other words, other paths may become critical paths with this enhancement. Figure 5.1 illustrates this concept.

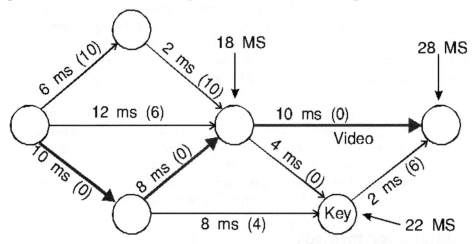

Figure 5.1 Enhancing a component that constitutes a bottleneck on a project may provide full or partial benefits, depending upon whether that affected path now has slack time. A 10 percent improvement on the path may reflect a 20-ms performance boost to that path—from 200 ms to 180 ms—and lower overall project completion time from 890 ms to 870 ms. This is only a 2.2 percent boost to overall project completion time.

Another impediment which affects me in most enterprise tuning projects is my inability to assimilate the enterprise design, NOSs, OSs, application software, integration ramifications, and assembly-line like operational process inherent in any computerized organization. I just cannot create a complete picture of the enterprise, read all the documentation for applications which are new to me, understand all the details for the software and development environments, and extract every optimization handle. Obviously, this is a general limitation of time and resources. I see this problem with other people too, not only those who are intimately familiar with the enterprise, but also outside consultants and vendors.

In fact, those who are most familiar with the enterprise have assimilated so many fine details that they cannot see the big picture. As a result, you will want to abstract the design of the enterprise and focus on the most urgent bottlenecks. Once focused, drive hard into the details, documentation, and the pragmatic solutions. Before haphazardly tuning a system, consider that there are always other options that may yield better results. You should know what results to expect from your efforts. Correlate that with the level of effort required to improve performance. You should answer the questions of why you need to tune the system if it is not broken, and who will provide time and materials for the tuning process.

Multiple solutions for the same problem represent a degradation of efficiency and performance, and dispersion of resources. For example, many large networks will have SNMP, hardware-based protocol analyzers, and hub- or router-based network management tools all running in parallel to gather essentially the same traffic and packet information. This wastes node horsepower, intermediate node resources, channel bandwidth, and management time. Another example that may be more frequent, but perhaps not as egregious, is to find Eclipse FAX, WinFaxPro, WinFaxPro network edition, Alcom network fax services, and a handful of Bitfax personal software applications running on the network. Although most of the fax traffic is localized, there is sufficient overhead for application storage, backup, management, and control related to these programs. Furthermore, every user must learn how to use or merely coexist with them. The most wasteful part of this duplication is that all of these applications are ultimately supported, debugged, and optimized. You cannot possibly know every aspect about everything on the enterprise network. Focus. Focus. Focus.

Optimizing with design

Faster equipment, faster transmission channels, and pricey consultancies cannot always solve performance problems. Some problems are solved with budgets, but many enterprise network bottlenecks are truly infrastructure and design issues more than hardware or device problems. You cannot just add lanes and new expressways to solve the social, distribution, and workload problems. LAN bottlenecks typically are solved with money. This is not so with enterprise infrastructures. Latency is usually the killer, while background traffic and delays from mixed protocols and synchronized traffic merely add to the gridlock.

Design is free. Better design costs the same as poorer design, but can yield substantial performance and management benefits. Implementing design is not always so free, but EIA/TIA 569/569 premise wiring standards can cut the move, add, and change costs for performance experimentation and any subsequent microsegmentation, rerouting, and backplaning. Furthermore, the hubs, routers, and switches are not cost-free. These expensive devices can, however, be used in better configurations to shorten latencies and contain background traffic. Selection of the correct routing protocols (for example, avoiding RIP or SAP over backbones) and network protocols also provide fundamental performance gains. The following vignettes show the fundamentals for isolating traffic, consolidating connectivity and routing, and replacing hobbled connections.

Figure 5.2 illustrates the most basic enterprise network with a spanning tree of gateways or routers. Although there are thirty-some nodes and two sets of alternate routing paths, this concept can be expanded to thousands of nodes and hundreds or thousands of routers. It matters very little which technology is in place, so long as packets are filtered and broadcasts are contained. The overhead differences between a true single-purpose router, and a server or workstation-type device functioning as a gateway and router, is immaterial. What does matter is that the traffic in aggregate will create performance delays and bottlenecks with this microsegmentation architecture.

Boundary routing (in place of bridging) and better routing algorithms can easily reduce background network traffic. This is important when multiple protocols are routed between domains, as with the central office and remotely networked sites. This vignette also shows how microsegmentation can be used to isolate protocol differences and traffic with certain characteristics. Performance is best when the servers and their clients are part of the same domains.

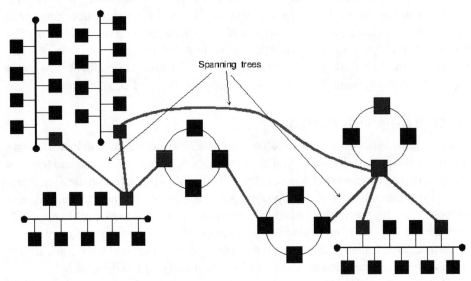

Figure 5.2 A network of subnets with the inefficiencies of spanning trees and multiple hops.

Consider how you can construct your microsegmentation domains to minimize internet traffic and bandwidth-eating broadcasts. Configure your routers to filter all broadcast traffic between domains, and upgrade software so that broadcast requests are resolved with less overhead. Domains also provide for secure firewalls in the event of a network panic. Figure 5.3 illustrates the establishment of network domains for the enterprise network environment.

Performance in terms of management overhead, domain configuration, adds, moves, and changes are improved with EIA/TIA 568/569 premise wiring concepts, or at least the implementation of a central wiring closet and wiring hubs. The hub minimizes the need for multiple router hops by establishing domain switching at the hierarchical levels, as Figure 5.4 shows.

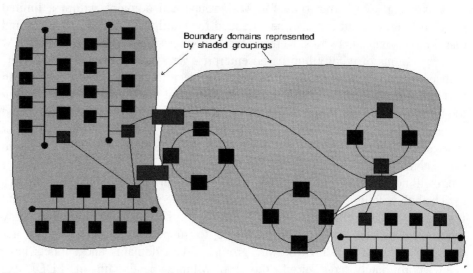

Figure 5.3 Boundary routing can contain local traffic from the enterprise network, and is an advisable first step for improving the performance of any backbones, backplanes, or wiring hubs saturated with background traffic.

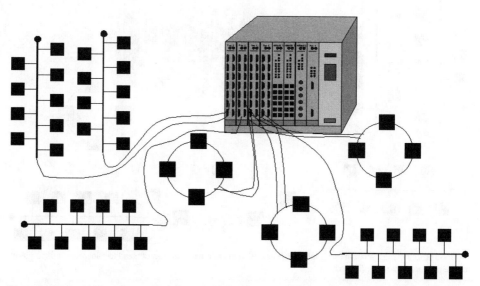

Figure 5.4 The multihop network is optimized by recabling connections into a hub and circumventing the complexity and latency of the multiprotocol routing and multiple router (gateways) hops.

This design works not only for workgroups and domains within a limited geographical area, but can also be extended for multiple floors in high-rises and distributed campus networks. Refer to the Synoptics slide show on the CD-ROM for illustrations of wholly hub-based enterprise networks. However, the performance of multiple hops and traffic routing between the constructed domains can be further optimized by reviewing the hierarchical configuration and creating loops in the design. Fewer hops mean better internetwork performance and reduced latency. The microsegmentation of the domains into smaller or reorganized subnets increases the available shared media bandwidth available to each subnet. This presupposes that the further reduction of domains into more of them will not increase the inter-domain traffic to saturation, as Figure 5.5 shows.

The hub with its 250 Mbits/s or greater backbone is not the only solution, and may not even be suitable for campus networks. The distance limitations may require construction of a fiber backbone to support 2 kilometer or greater distances between buildings or workgroups. When domains and subnets have been appropriately configured, the 100 Mbits/s bandwidth of FDDI can adequately provide E-mail, host, and client/server support on the enterprise network, as shown in Figure 5.6. You can also create more complex structures with hubs and FDDI backbones. However, not even FDDI, ATM, or any current protocol will handle the cross-traffic from verbose LANs.

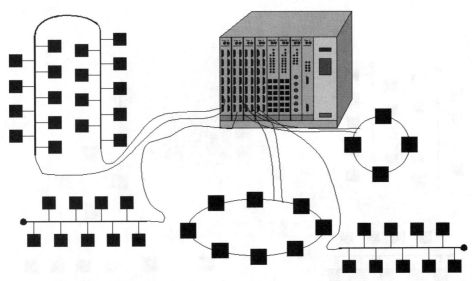

Figure 5.5 Subnets (Ethernet in upper left and Token-Ring in bottom center) reorganized to minimize interspan traffic for the multiprotocol router (in the hub).

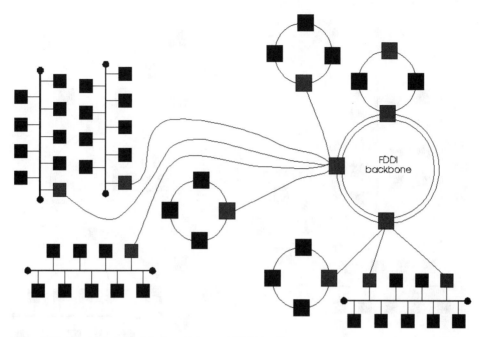

Figure 5.6 A network of subnets linked to a high-speed (FDDI) backbone.

Although dual homing is primarily a feature of FDDI, it can be enabled with NetWare Balance NLM and other network operating systems. Furthermore, dual-pathed traffic is not only a function of FDDI, but could be used with most other LAN protocols, and even with WAN connections, as detailed in Figure 5.7. The performance benefit is mostly a feature of doubling (or tripling, etc.) the shared media bandwidth. (Redundancy is often an important network criterion as well, and is provided with dual homing or bandwidth-on-demand configurations.)

Also note that broadband WAN connections need not be symmetrically supported. An outward-bound connection can be established on 56 Kbits/s PVC but the inbound traffic can be routed through an on-demand 9.6 dial-up modem connection; this is less expensive than maintaining dual higher-speed circuits. Multiplexers (muxes) or inverse multiplexers (I-muxes) also allocate directional bandwidth and piggyback signals over the primary transmission channel.

Several types of multiplexers are in common use in the long-distance carrier market and support dedicated or switched circuits. Time division multiplexing (TDM) integrates multiple channels into different time slots in the transmission channel. This is inefficient if the different traffic streams are dynamic or are idle much of the time. The inverse multiplexer provides automatic selection of frac-tional channels for data or voice communication, or it can also add circuits as needed to support dynamic bandwidth requirements. The use of I-muxes with drop-and-insert connections will *not* support bandwidth-on-demand or any dy-namic bandwidth allocation.

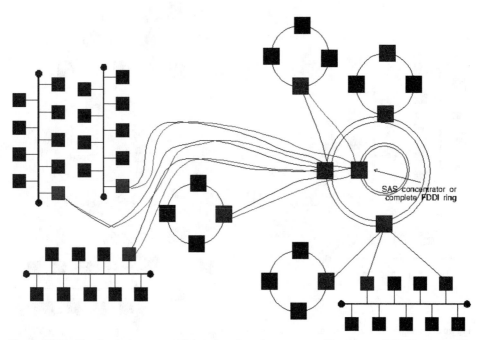

Figure 5.7 Dual-homing and more sophisticated configurations are possible with an FDDI backbone. The configuration is not only a performance enhancement in that rings that are properly load-balanced can carry about twice the traffic load, but failure of a ring and counter-ring is protected by the secondary paths.

Cell switching, statistical multiplexing, and inverse multiplexing represent other applications of this same technology. Cell switching, better known for use in frame relay or ATM requires header and management overhead. Statistical multiplexing is often used with LAN traffic because it can allocate bandwidth as needed. However, it also requires header and management overhead. Inverse multiplexing is the opposite of TDM in that it is used to combine multiple DS0 or T-1 lines to avoid the higher costs of DS-3.

Primarily, muxes do not reflect performance optimization but rather the financial cost reductions for leased lines, T-1, and dedicated lines. As information transmission bandwidth requirements have increased, primarily with regard to remote LAN interconnectivity, the technology has been applied to extend the available bandwidth. Although I assert this is also primarily a financial optimization, increasingly multiplexing is pertinent to resolve the fundamental bandwidth limitations of available WAN links. For example, sometimes a router will not sustain more than a single point-to-point connection. Since the largest available link might be a T-1 line, multiplexing multiple streams into this channel represents the only means to provide more capacity.

When FDDI seems limited in bandwidth, even it can be collapsed into a backplane for speeds ranging from 250 Mbits/s to 10 Gbits/s. The design limitations of microsegmentation, domain configuration, and internet traffic still remain in force. The backplane is still a shared media environment, but the shared media is 2 to 100 times greater in capacity. Because of this, the backplane can provide greater routing capacity and connectivity between segments, subnets, and domains, as Figure 5.8 illustrates. This configuration could also support dual homing where needed for mission-critical applications.

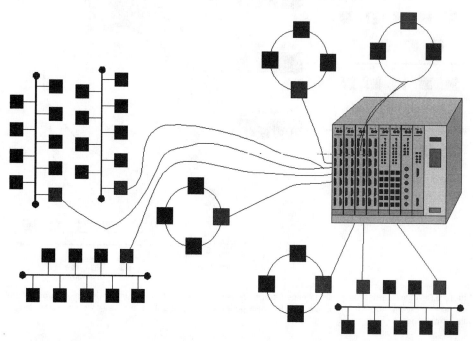

Figure 5.8 A backbone replaced with a backplane in a hub providing greater bandwidth than FDDI.

The limitation of the wiring concentrator or hub is, of course, its ability to drive a signal past 100 meters (usually less for other protocols and Ethernet variants). By the way, just because Token-Ring can support 100-meter lobes, remember that the networks must be built to conform within the adjusted ring-length limitation for segment bypassing and ring wrap. Although hubs can support fiber-based FDDI, be careful that you do not degrade the benefit of the high-speed backplane by piping high-speed traffic to slower distributed links. In fact, most multiple-floor environments are very likely to need judicious microsegmentation in order to conform to lobe wiring length limitations. When this site supports virtualized subnetting, as is becoming more prevalent with enterprise networks, the design is quite sensitive.

Specifically, the design should support bridging, routing, or switching at the LAN-level, have a layer for LAN interconnection, and provide external (peripheral) routing services for backbone access, as shown in Figure 5.9. As stated earlier in the chapter, the benefits of virtual switching are lost when traffic is routed over any shared media segments. You might try to avoid this limitation by keeping these segments and subnets on the periphery of the core enterprise network. The shared media sections can be terminal subnets, as stated in "Virtual switching" on page 223.

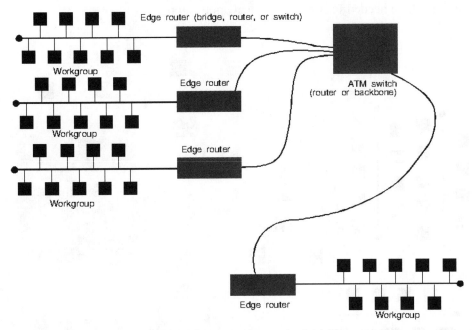

Figure 5.9 Virtualized networking or extended corporate sites require that LAN connectivity be separated from peripheral backbone access. This is not only for strict performance optimization, but also to support lobe length wiring limitations.

Although I refer to the EIA/TIA 568/569 premise wiring recommendations, Figure 5.10 shows how you would actually wire nodes into a central wiring closet and thence into the hub. Jumpers or pigtails patch each lobe cable into the hub, or a wiring octopus (called a "hydra") can minimize the need for spaghetti wiring. Not only does this simplify the adds, moves, and changes necessary in the enterprise network, it also simplifies the virtualization of domain configuration.

However, this architecture has no relevant effect on throughput or latency, as it does not alter transmission levels, signaling speed, or anything else other than managerial or janitorial performance. The performance differentiation between the Ethernet bus, 10Base-T, 100Base-T, and Token-Rings is no longer important at the physical wiring level. They all perform within the traditional limitations of LAN protocols. (100BaseVG-AnyLAN is a different wiring and protocol scheme altogether.) However, these are typically easier to manage at the physical level if you migrate toward TIA/EIA category 5 copper twisted-pair. Matrix wiring and Category 5 wiring provides a scalability solution. The matrix wiring hub can actually construct the backbone segments. This improves management performance, but also provides significant freedom to reconfigure the network for better performance between domains.

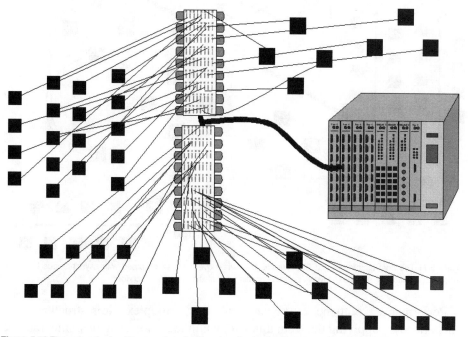

Figure 5.10 The enterprise backbone replaced with a virtualized network and premise wiring.

The mainframe or other host throws a wrench into enterprise network performance. Because SNA and SDLC are synchronous, performance and LAN-type traffic is likely to be compromised. Multiple hops and routing problems merely compound the performance bottlenecks, as Figure 5.11 illustrates.

Dedicated router-encapsulation and traffic prioritization by protocol and traffic type is not reliable. Hub-based FEP connections with switching virtualization is a distinct performance improvement, and also lowers the cost for making connections between hosts and typical LAN equipment and workstations, as Figure 5.12 shows.

The reintegration of host, LAN, and client/server activities into the enterprise infrastructure usually represents a primitive attempt to provide all services to all sites and users. It works until the typical performance bottlenecks saturate the connecting channels and claim application latency victims. This hodge-podge integration is best reflected in Figure 5.13.

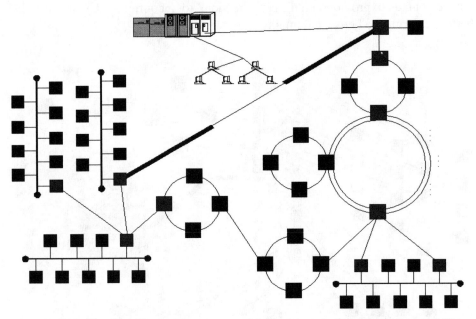

Figure 5.11 The host mainframe and FEP creates bandwidth overloads when piped with asynchronous LAN-type traffic and potentially overloads gateways providing routing and traffic translation services.

Performance bottlenecks, even for this complex infrastructure, are ameliorated by applying the techniques shown in the prior vignettes. Integrate the infrastructure and install backplane centralization or virtual switches. If the performance is needed and you can adapt to the longer latencies, establish appropriate domains for segments of the enterprise network. Even hosts can be integrated into the enterprise as part of their own private domains. You might want to establish secondary connections to remote sites for client/server or host traffic to minimize the traffic on the enterprise network backbone. But, notice in Figure 5.14, the enterprise network is not necessarily a singular entity with a singular transmission channel. The semistar wiring topology obscures the configuration of the domains (segments and subnets) and the multiple paths available through the high-speed hub-based backplanes.

By the way, one of the side benefits of premise wiring integration and consolidation of equipment into hubs and high-speed backplanes is that peripheral connectivity devices (such as gateways, routers, bridges, and switching devices) are no longer needed. The centralized hubs consolidate the services.

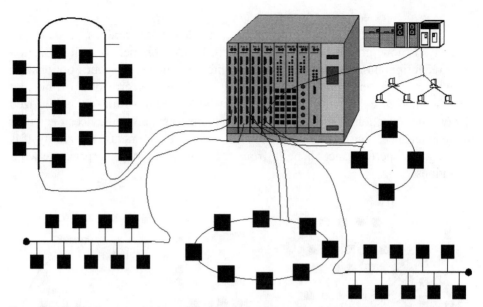

Figure 5.12 Two solutions for mainframe synchronous traffic include partitioning the enterprise network backbone for LAN and host (and voice) traffic, or offloading the FEP and gateway with a hub-based adapter product (such as the Synoptics processor as shown on the CD-ROM).

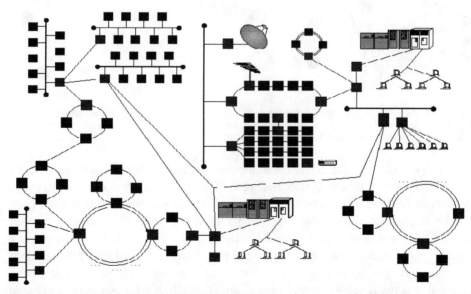

Figure 5.13 Performance bottlenecks on the enterprise network are caused by heterogeneity, wide-area geography, and the difficulty in managing both a dispersed and multiple-platform architecture.

Additionally, the microsegmentation provided with bridges, routers, or gateways is now provided at the hubs as well. In addition, network virtualization and reconfiguration with hub-based switching increase your options to reconfigure the enterprise on the fly, or tune performance ad-hoc; a microsegmented connection could exist just long enough to transfer a digitized X-ray. The reduction of the number of peripheral connectivity devices decreases the maintenance, configuration, and overall energy requirements. This in itself is a substantial performance benefit from the standpoint of enterprise network operations.

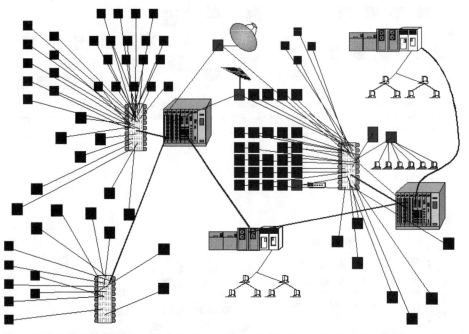

Figure 5.14 When the enterprise network contains multiple LAN subnets and remote sites, ultimately performance increase is a function of minimizing the workload routed over hops, consolidation of services, faster transmission channels, construction of premise wiring and network virtualization.

Finally, optimize the connection method. Remote sites do not need to be connected with a deterministic connection. Although bandwidth is often the bottleneck for WAN linkages, the connection is two-way and the bandwidth need not be equilateral. You could have faster linkages outbound and slower ones inbound. Furthermore, you can optimize the bandwidth to meet the burst or peak loading on a directional basis. In addition, consider bandwidth-on-demand or different connections to improve performance. Do not think in a single plane. Enterprise network design is multidimensional. While it may seem very desirable to minimize the number of LAN protocols, transmission technologies, and hardware by moving to a uniform platform, environment, and management methodology, WAN connections are expensive and by no means uniform in cost based on load, service, or speed. You can optimize costs, performance, and latencies at the same time by designing linkages to match performance, as Figure 5.15 illustrates.

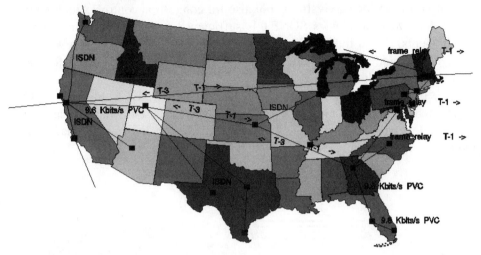

Figure 5.15 Maximize WAN performance with mixed-speed transmission services; minimize costs by selecting the most effective and efficient protocol to traffic composition.

There are no rules that state you must provide uniformly bidirectional services. You can mix and match not only the protocols, but also the connection type, the service providers, and the CIR. You may find that there are irreconcilable discrepancies between tariff rates, service guarantees, and usage-based plans. Pricing isn't the only issue, as availability and reliability are also important concerns; these can be addressed with Make Systems Analyzer, as described in Chapter 4, or other tarriff optimization products. You are likely to discover that adding bandwidth for remote connections could take longer than stringing new wires internally; it could take weeks. When thousands of dollars are at stake, assess the options. However, before committing to a network of mixed media, vendors, protocols, CIRs, and services, also factor in the management, training, and headache costs long-term for this heterogeneity.

Also, recall from Chapter 3 that typical WAN connections are rated with a bidirectional channel capacity, so that 9.6 Kbits/s means that the overall throughput for *both* directions must be less than the rate. Overhead, synchronization, compression, in-band management, and signal blocking reduces that capacity. When you build your WAN performance simulation model, factor in this very important limitation. However, realize that frame relay, ATM, or ISDN often support bidirectional throughput at the rated capacity for approximately double capacity. In particular, it is easy to overlook that ISDN supports two D channels which can be multiplexed together or dedicated to full bandwidth directional support.

It is important to realize that ATM runs on top of SONET in either a unidirectional or path-switched and bidirectional architecture. This is important not only for performance optimization and enterprise network design, but also for reliability and survivability. For example, when SONET is combined with path protection switching (PPS), the topology includes a virtual ring. This is useful not only for ensuring survivability, but also for congestion control. Although you might think that 1.6 Gbits/s SONET is sufficient for most any communication requirements, if past traffic growth is any example for the future, you will need greater bandwidth and adequate congestion control mechanisms even for these enormous bandwidths.

Path protection switching provides duplication of SONET (and hence, ATM) signals transmitted simultaneously over unique paths, the best of which is selected for delivery. In Figure 5.16, node A is the source and node B is the destination, and multiple virtual rings ensure reliable delivery and speedy transmission.

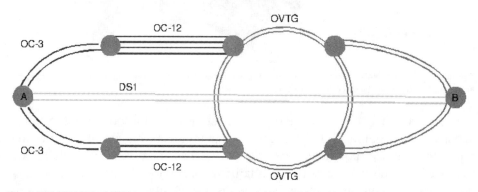

Figure 5.16 SONET (and ATM) benefit from path protection switching, which enables virtual rings.

Furthermore, virtual rings are not only for high-end technologies such as ATM and SONET. It is also easy to construct a virtual ring for congestion relief with standard networking components, as Figure 5.17 indicates. Extra traffic cannot be handled with the link between A and B operating at capacity. By rerouting overcapacity through node C, the ring between nodes A, B, and C are switched (as with a spanning tree or hub-switched environment) to support additional traffic. It is probably pertinent to upgrade the capacity of the route directly between nodes A and B, however, to address the chokepoint more directly.

Virtual rings are more flexible when overlaid on a framework of path-switched rings. Although this is primarily a SONET concept, the IEEE 801.10d spanning tree specification makes this feasible for both local and WAN segments of the enterprise network. Mesh ring configurations push this technology further, providing multiple interconnected paths. Figure 5.18 illustrates the utility of redundant switched connections for increased reliability and performance.

The actual architecture is usually dictated by the type of network involved, local exchange carriers, and the economics associated with each, as well as specific sets of customer applications. An access network is a "homing" configuration, in which most of the traffic from multiple locations zeroes in on a single, central office location. This matches the configuration of a corporation with a central site and many remote offices. Generally, local exchange carriers are putting unidirectional path-switched synchronous-optical-network rings on the access side of their networks.

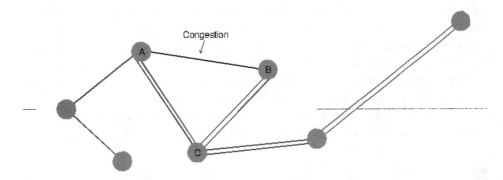

Figure 5.17 Virtual rings provide a means to route or switch past congested channels.

As stated earlier in this chapter (see "Bandwidth on demand" on page 214), the bandwidth in a path-switched ring must be dedicated all the way from transmission site to receiving site for each signal. Because bandwidth is not reused in a homing network configuration, path-switched rings are more economical on the access side.

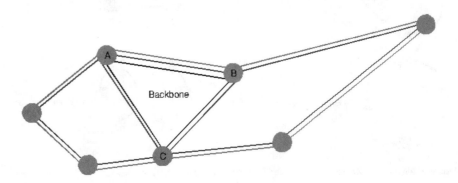

Figure 5.18 Multiple virtual rings demonstrate a meshed-ring configuration.

In contrast to the access network's homing configuration, an inter-central office network usually is a "mesh" or "star" configuration, in which there is uniform distribution of traffic between and among multiple central office locations. If one central office goes down, a carrier needs to reuse that fiber to reroute the signals; otherwise, unused bandwidth is wasted. Bidirectional line-switched ring (BLSR) is the normal configuration. Other common configurations include point-to-point (ADM), DACS hub, and mesh DACS. The common configurations are pictured in Figure 5.19.

Based on SONET platforms that usually run at OC-3 to OC-12 rates (155 to 622 Mbits/s), a path-switched ring transmits a signal in opposite directions around the ring, on two separate channels within the same fiber. The switch at the receiving node looks at each and uses the higher-quality version, or the first arrival if it is of sufficient quality. In addition, path-switched rings on the access side let carriers offer even greater service protection, because they allow for dual feeding—dropping a signal to two separate central offices. This is dual-homing (an FDDI configuration) for WANs. Therefore, even if a disaster knocks out one central office, the other takes over to ensure that the customer does not lose service. Path-switched rings offer a choice between OC rates on the access side. Financial institutions, design teams, simulation, forecasting, and possibly even video may justify your selection of the OC-12 rate in a loop ring.

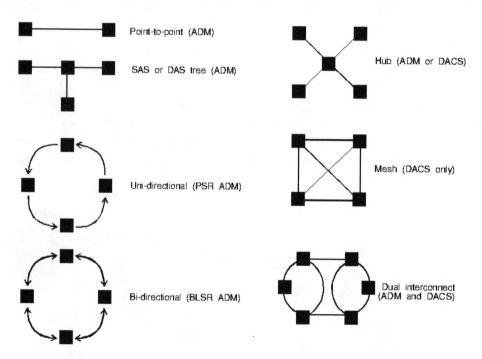

Figure 5.19 ATM (on SONET) architectures for performance and survivability optimization.

Along with long distance carriers, most local exchange carriers are deploying bidirectional line-switched rings from within central-office networks. A BLSR, based on platforms running at the OC-48 (2.4 Gbits/s) rate and higher, also transmits a duplicate signal in opposite directions around the ring. However, it uses 24 of the 48 time slots on one fiber to serve as the working ring, and 24 time slots on the other fiber to serve as the protection ring. If one signal fails, the node requests a signal loop-back around the failure. The signal loop-back in a BLSR, which allows for reuse of bandwidth, makes this architecture more cost-effective in the high-speed, high-capacity inter-central-office network. Depending on traffic patterns, a meshed inter-central-office BLSR network can support up to one-and-a-half times the amount of traffic supported by a path-switched ring.

Generalized optimization techniques

The most effective technique—bar none—for optimizing overall enterprise network performance is to minimize the workload on the entire system or on individual segments and subnets of the network. You will find it easier to reduce the load on the system as a whole than to make a current set of tasks run significantly faster. Stop background tasks. Run batch operations after peak hours when traffic loads are lower. Provide alternative paths—secondary routes for high-volume. Hire a person to perform data backup early in the morning, late at night, or on weekends to off-load this traffic from the network. Consider moving multitasked operations to another segment. When all else fails, remove all trivial applications and .DLLs from the network and load them from local hard drives.

Your success rate will be higher with removing traffic and simplifying the environment than with trying to increase its throughput. Recall the paradigm of the Wall Street layoffs from Chapter 2. Performance is an issue of the infrastructure, and changes to that infrastructure have more effect than segment or intersection modifications on the enterprise network. Design of the enterprise network is the bottleneck you need to resolve.

Designing the enterprise network

More than any other aspect, design and configuration of the actual infrastructure has a more profound effect on enterprise network performance. It surpasses bandwidth issues, the selection of specific protocols or connectivity boxes, or the applications actually run and the traffic levels required. Design should encompass the following items:

- Purpose
- Redundancy
- Integration
- Wiring infrastructure
- Security
- Scalability

Design is a big-picture item, and the enterprise network is also a big-picture item. Before you even consider addressing specific bottlenecks, review the enterprise from an overview. Although design does include the issues of Ethernet versus FDDI, it is much more than that. The six key infrastructure topics are expanded in the next sections.

Purpose

The purpose for the enterprise network is usually driven by top management directives or the underlying focus or purpose of the organization it is to serve. You need to understand the organization and its function before beginning optimization. You want to understand why some things must be done and why others are immaterial. You certainly do not want to expend the effort to maximize performance for operations that are not of consequence.

In many cases, the primary purpose of the enterprise network is to integrate communications for the organization. In some cases, the enterprise network is a downsized replacement for a mainframe data processing environment, and its purpose is to reduce MIS costs. When advancing technology is the primary purpose of the enterprise network, optimization may mean thorough access to resources and full interconnectivity and interoperability. Whatever the purpose for the enterprise network, make sure that you know what it is and that you design performance optimization and tuning projects in compliance with it.

Wiring infrastructure

The infrastructure is the basic framework for the enterprise network. It is the foundation for all that follows, much as the iron girders define the shape of the structure in Figure 5.20.

Read the EIA/TIA 568/569 premise wiring recommendations and addenda; they explain the concepts and goals for the hardware infrastructure for any enterprise network. Although most enterprise networks result from the growth of LANs and interconnection of WANs, and thus the wiring infrastructure already exists, you should understand what flexibility and options the centralized wiring provides. Even if you cannot consider the costs for conforming and upgrading to the EIA/TIA recommendations, at least learn the concepts for centralized wiring that make possible a flexible, configurable, and scalable infrastructure.

For example, you can view your IBM Type 1 Token-Ring wiring as sunk assets without any value. They seem useless because they are used and stuck in the ceiling, thus seemingly not recyclable. Somehow, they are different from EIA/TIA Category 4 wiring—so says your sales rep who wants to pull them from the ceiling to make way for new wire that conforms to the specification. However, wires pulled through ceiling plenums from a 110 punchdown block to individual outlets in offices look just like a premise wiring system; it is all a matter of visualizing what you have available to work with. Of course, if you can certify the installed IBM Type 1 wire using a ring or cable scanner as Category 4, you are significantly ahead in terms of network optimization. You now have a valuable resource that you can reconfigure and recycle into your new structured wiring facility. This is all in the perception. It certainly helps if you have logical schematics of what the network does, how it logically interconnects, and how it is physically wired, as Chapter 4 demonstrated.

Figure 5.20 The wiring is the underlying framework for the enterprise network.

The wiring infrastructure must reflect the environment, geography, weather, and expansion needs. It may need to incorporate wiring legacies as well, and the point-of-presence limitations of a long-distance commercial carrier.

Redundancy

The enterprise network needs redundancy built into the infrastructure. This is a performance issue. Because there are apt to be so many traffic intersections and operations critically dependent on services, the failure of these points could spawn a network panic. Such an ultimate performance failure spreads far beyond the cause of the problem and the result is way out of proportion to what it should be. By definition in this book, the enterprise network should provide multiple routing paths, and you should want to design these paths so that failures are bypassed. While the need for redundancy might seem obvious for connectivity devices, you will also need this for file servers, hosts, and anything supporting compound documents. Services, including database processing, SQL fulfillment, printing, and files can be backed up by tandem hot sites or through remote HRM.

One concept to note is that the enterprise infrastructure itself does not only need redundancy to ensure adequate levels of performance, it can also provide it as a service through its WAN and remote links. Hot sites can be linked to operational environments at all times through the enterprise network for continuously updated remote archival storage and instant readiness.

Security

Enterprise network security is difficult to add as an afterthought. If security is important, it must be designed into the infrastructure. Fiber can replace copper wire. Copper twisted pair can replace coaxial cable. Private virtual circuits (PVC) can be selected instead of a local exchange carrier facility, and connectivity devices can be selected that incorporate firewalls, dial-back, authentication, encryption, compression, and remote SNMP monitoring. Because the physical enterprise infrastructure provides the logical access to virtually any device on it anywhere from anywhere, security is critical.

You certainly may not want to minimize enterprise networking by detaching operations with sensitive data or privileged applications, and thus providing limited access to it. Although that is a viable and not unreasonable solution, it is contrary to the concept of global access to information. There are other reasons for limiting access—including optimizing performance by reducing non-priority traffic—and these are viable and reasonable too.

Integration

Network requirements change. Three years ago, the R&D LAN could have represented cutting technology for an organization. Today, that same LAN has probably grown four or five-fold as demographic information details. That one LAN is paralleled by one for sales, one for marketing, one for documentation and publishing operations, one for legal, another for accounting, many supporting the traditional office activities, and one or more each at each remote site. Finance has probably dumped most of the 3270 terminals for cheaper and more flexible PCs and links into the host mainframe through a DCA IRMAlink gateway, or something else like it.

Enterprise network integration is rather a case of reintegration. LAN, client/ server, and mainframe operations diverged for most organizations and took very different management and operational tracks. Now that the infrastructure can support these technologies in tandem, it is important to think about integration as a cost-effective consolidation, as a means to bring telecommunications and data communications management under control, and as a technique for optimizing overall organization informational performance.

Because you want to include every possible computer with all the diversity on the enterprise network—that is a goal for many network managers—integration is a first-line design concern and constraint. Although not every computer should attach directly to a backbone, ultimately the hierarchical infrastructure must be able to support routing, interconnection, and interoperation. As many vendors try to convince designers and managers that ATM is the solution for every bottleneck and the uniform infrastructure for LAN, WAN, and enterprise, the importance of integration in the design will be all too apparent. Specifically, there are products currently available with three different implementations of congestion control and flow control, best known as available bit rate. These controls prioritize voice, video, and data bursts.[1] These methods seem to be implemented a little differently by each vendor.

Integration is a constraining issue for WAN connections. Although there are multiple players for local exchange carrier services—AT&T, Sprint, MCI, WilTel, BT, TTI, BTI, and all the local Baby Bells—you probably want to minimize the number of vendors but still have redundancy and a uniform transmission environment. Having too many protocols is expensive in terms of equipment, maintenance, and management. However, many local exchange carriers do not have a complete point-of-presence for all telecommunication services.

Furthermore, while vendors may be able to standardize congestion control for the enterprise network switches, your remote traffic will typically travel through a cloud of services provided by one or more local exchange carriers. Your priorities and committed information rate agreements may not truly map to how the technology is implemented in the carriers' switching network. It is also not clear how they will handle cloud congestion in any event. By the way, the same congestion issues are relevant for X.25, frame relay, and SMDS.

In-house technical support and help desk facilities are generally overburdened with supporting Word for Windows, MacWrite, WordPerfect, AMI, and probably several other word processing legacies. E-mail is usually the most insidious end-user integration problem. The issue for integrating and consolidating these separate operations is how to rectify the many choices for hardware, software, and operating systems. LAN Manager or LAN Server compete with Vines, NetWare, Sun NFS, DECnet, Pathworks, and a few other network operating systems. The same holds true for this more complex software, only more so. Heterogeneity must give way to homogeneity. Figure 5.21 illustrates the percentage of the four largest LAN operating systems integrated into corporate enterprise networks.[2]

While choice or a laissez-faire attitude might seem politically astute for LAN evolution and deployment, it undermines enterprise network optimization at all levels. This includes disk space, backup, network loading, bandwidth, latencies, network management, troubleshooting, user support, and reliability. The big picture cannot include a highway with users driving to their own individual

---

[1]*ATM Forum Ponders Congestion Control Options*, Stephen Saunders, *Data Communications*, Pp. 55–60, March 1994.
[2]Business Research Group, Newton, MA, 1994.

traffic rules. That represents chaos, which has a way of spreading beyond even LANs that are fully encased with firewalls. After all, E-mail must be delivered.

Although protocols, products, vendors, and software differ, standardized consolidation provides a fundamental opportunity for optimizing network operations and performance. So what if a group thinks one vendor provides a better router than another? It matters most that the infrastructure conforms to a simple and singular metaphor. Consider what might happen if you decide that one router is in fact better because it provides dedicated bandwidth switching. The technology always changes, and something always will be better. The infrastructure—from the wiring to the hub that consolidates all the patch cords—is unified and the exchange of one router with another—albeit in multiples—is technically feasible, controllable, and quickly accomplished. This may be expensive because it is based on multiple replacements, although it will be less costly and more reliable than maintaining routers from multiple vendors.

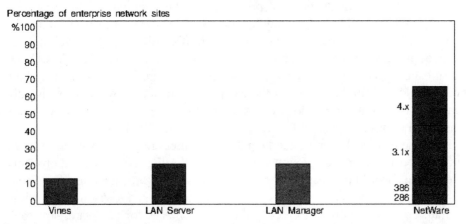

Figure 5.21 The percentage of corporate enterprise networks running common NOSs.

Integration may also imply the need for standardization. Multiple LANs with various media protocols, transmission speeds, routing protocols, and proprietary operating systems represent a complex puzzle that is sometimes best simplified. This is not always possible immediately when integrating preexisting hosts, LANs, and operations. Integration probably needs to reflect legacy operating environments. However, higher long-term, management, and maintenance costs often create a significant reason to optimize by standardizing.

Scalability

The need to increase services on enterprise networks is a significant design issue because demands are likely to outstrip capacity by two, three, and four times within the next few years. Refer to Figure 3.15, which shows where network managers expect bandwidth utilization growth within the next year. Imaging, graphics, and more of the same top the list. This expected growth in network services will require a corresponding increase in the infrastructure. This includes the wiring plant, the servers, the number of users on the enterprise, and the connectivity devices. The obvious solution is to scale up the infrastructure.

However, the point is not merely to double or triple the infrastructure to support the traffic demand, but create an infrastructure that is inherently scalable so that you can add bandwidth, processing power, and services without redesigning it. For example, a digital broadband PVC providing 9.6 Kbits/s is scalable up to 56 Kbits/s (6 times greater). This link is not particularly flexible, and given most bandwidth requirements represents the choice of a poor infrastructure. However, if you change the infrastructure so that the link is established with frame relay (at 56 Kbits/s), you have the option to scale up that link to T-3 speeds (108 times more) without making software, hardware, or wiring changes.

There is another reason for designing networks so that they are scalable. First, they are inherently manageable when a technology is homogeneously applied. It is as easy to manage the 56 Kbits/s frame relay as it is to manage the T-3 (or E-3) connection. Second, DP work is increasingly performed cooperatively or collaboratively. Such work groups are dynamic, and scalable architectures inherently provide the flexibility to duplicate resources or reconfigure a portion of the network to match these demands. Dynamic switching technology is particularly relevant as a component in the enterprise network infrastructure.

Financial optimization

There are many aspects to enterprise performance optimization that include more than just pumping through traffic as quickly as possible, or replacing antique components with faster ones. Not all performance solutions are technical; some are financial. Costs are balanced out by budget, management considerations, strategies, and economies-of-scale. The following list shows the issues discussed in subsequent sections:

- Consolidation of duplicate resources
- Consolidation of transmission services
- Better productivity from people with better communications
- Decreased use of fax and meetings, and more reliance on networking
- Optimized transmission costs
- Overlooked benefits of performance optimization
- Used equipment

Consolidation of duplicate resources

Consolidation is the leading effort for financial optimization of enterprise networks. Duplicate (and underutilized) servers, routers, printers, and disk space represent not only a management and operational headache for the network management organization, but also yield a poor return on investment (ROI). In fact, if you perform not only a physical audit but also a process review, you may find duplicate efforts and redundancy. Obviously, there are some fine points of politics, departmental ownership, etiquette, and friendships at stake when you suggest consolidation. Host operations are prime candidates for rightsizing with LAN-based stand-alone software and client/server applications. Also, consolidation of protocols and homogenizing the environment goes a long way to minimizing costs and maximizing performance.

Consolidation of multiple servers with a single superserver yields immediate savings, not only in terms of maintenance and support activities, but also in terms of the number of software licenses required for the NOS. Novell, Vines, and most other vendors provide licenses per server (and per site). As such, it is almost always less expensive to provide client/server support with a single superserver than five or more parallel servers.

Consolidation of transmission services

Consolidation of services is not limited only to data communication equipment and the networking environment. All communications can be integrated into the same carrier channel by combining voice, data, and video into single lines. These create effective economies of scale, because a larger channel is less costly than many smaller ones. Multiplexing is not new, but seems to be the newest of the "new" technologies for the LAN management groups. It is particularly effective for WAN links, which have not only sunk installation and ongoing fixed monthly costs, but also usage charges. LAN infrastructures are usually accounted for with different mathematics. Data compression can also extend the economy of WAN links. It is particularly relevant when you are pushing the capacity of a line and need to add more capacity. You can bypass some of the additional sunk, fixed, and variable costs by compressing transmissions at the end points. Expect about a 2 for 1 improvement in throughput with minimal latency increases. Vendors claim as much as 7-fold improvements, but the compression algorithms typically provide 1.8:1 to 2.1:1 compression ratios on standard network data. CVSD yields voice within 8 Kbits/s, while LPC requires only 2.4 Kbits/s.

Better productivity from people with better communications

Clearly, slow transmission slows productivity. You will have data entry people waiting on the host and transmission channels, people twiddling their thumbs waiting for the server to reply to a data request, and network panics ceasing all activity. Optimization does mean balancing the costs of improving the network against the benefits of increased productivity. But also, financial performance of the enterprise network can be tuned by improving the distribution of and access to information. Getting facts and information to the correct people increases productivity and gives them the means to make better decisions.

Decreased use of fax and meetings, and more reliance on networking

Voice chat, E-mail, internal distribution lists, and networked bulletin boards are more efficient than voice mail, multiple faxes, and face-to-face meetings. Integration of mail services and communication has significant benefit on large enterprises. However, the problems of getting cc:Mail, Alcom fax services, DaVinci mail, and the many other time and billing or calendaring programs integrated is a VAR nightmare. Yet, in spite of the difficulty with these existing technologies, telephony on the enterprise with TAPI (telephony API), networked voice mail, white board, and videoconferencing calls is just beginning.

The financial benefits of voice on data are in the integration of infrastructure and reduction of duplicated services. Telephony Services for NetWare provide click and call from a centralized enterprise network phone directory and database, drag-and-drop conference calls, on-screen integration of voice mail, E-mail, and faxes with the same mailbox metaphor and interface, and interactive voice response for orders, technical support, and other services with DID touch-tone codes. This is not only useful for automating the organization business, but also for managing internal network activities. The same services are expected in Microsoft Windows, Daytona and Chicago (Windows 95), releases during 1995.

Fax service optimization
There are inbound routing schemes. These include direct-inward dial (DID), dual-tone multifrequency (DTMF), channel, and colon and source identification routing. DID is a popular method that requires special phone services and equipment. A single DID trunk line, obtained from the telecommunications pro-vider, can typically support 100 unique telephone numbers. This allows many users to have their own fax number. When faxes are sent to a specific phone number, the server matches it to a user and then routes the fax on the enterprise.

DTMF routing, another popular method, employs push-button phone service, allowing fax senders to enter an extension number after they have dialed the fax number for direction to a particular user. Channel routing, for instance, directs all faxes received on one channel to a single person or a group. The source ID method routes faxes based on the customer subscriber identification number (CSID) of the fax machine or server being used to send the data. All faxes that are received from a specific fax machine can, for instance, be routed to a certain person or group. Intel uses an approach called colon routing which only works by decoding the recipient's NetWare logon name to the incoming image. Others use the Group III header information to communicate the name of the fax recipients.

Optimize transmission costs
Transmission of data, voice, video, and other digitized or analog data is neither free nor priced at a fixed cost. Quality of the line is also an important factor. Where available, order a *data-quality* analog phone line or shift to switched digital services. This is important primarily for WAN connections, but even for campus networks, reliance upon a local exchange carrier or PVC for connections, overflow, or capacity on demand are suitable services. Price per circuit and price per MB vary widely based on capacity, service provider, and speed. There are break-even points that do make one service a better financial and technical match than others for your environment. For a financial analysis of transmission

methods, see "Finally, optimize the connection method" on page 190 for some pro forma analysis of carrier charges.

Another approach for sending thousands of faxes per day is to route them through a fax packet assembler/disassembler (faxPAD) and then over public data networks. This bypasses the long-distance services. This is feasible with in-house time-division muliplexing (TDM) and X.25 networks. Additionally, this conversion can also be buffered for transmission during off-peak hours.

Overlooked benefits of performance optimization

Savings from staff reductions are probably the most obvious benefits that can come from integrating enterprise networks with existing information systems, or a transition to new technology such as client/server applications or sales force automation. Other tangible benefits from new technologies often go unrecognized in the kind of standard ROI analysis used in most corporations. For example, in making the decision on enterprise networking, several tangible savings could result from a new system. These can include:

- Reduction in the need for (and costs for) new equipment.
- Increasing the speed at which dunning notices are sent to late payers.
- Shortening the monthly general ledger closing cycle.
- Performing "what-if " analysis during the financial-planning cycle.
- Reducing system support costs for existing mainframe-based, in-house accounting systems.
- Improving progress billing efficiencies by issuing progress billings three weeks sooner, and by reducing errors in billings.
- Reducing the time (and cost) of preparing budgets, business plans, and proposals due to increased availability of business data in real-time.
- Reducing the cost of generating financial statements.

Recycled equipment

Consider optimizing enterprise network expansion costs by recycling or buying used equipment (at 30 to 80 percent less than list). An Ethernet network built in 1994 functions the same as one built in 1984. Although the older devices may not be SNMP-compliant or remotely manageable, they still conform to the same specifications and should work well. So long as reused equipment conforms to the big-picture concept of premise wiring, homogeneity, scalability, and easy integration into the existing network, use it! There is a thriving market for used telecommunications and data communications equipment and support services.

With this same logic, recycle or resell your old equipment. If you replace your existing routers with all new ones from a new vendor, the old equipment is probably not scrap. Another organization may be trying to integrate and consolidate their burgeoning enterprise and think your old routers, for example, are excellent. The price is probably very competitive too. When analyzing your enterprise network optimization, factor in the value derived from reselling discarded software, hardware, licenses, site-licenses, and existing support agreements. While it may seem immediately expensive to consolidate and homogenize the environment, the resale could even recoup all those expenses and provide a long-term administrative and management savings as well.

Technical solutions

Here is a brief history of LAN (and enterprise network) optimization. It is useful because the tools have changed and some of the techniques now have formal names. Initially, computers were attached to LANs for resource-sharing. This changed as PCs became viable platforms with usable tools, and hence viable development environments. Initially, as Chapter 3 explained, the network bandwidth created by PCs and early engineering workstations was so minimal that the primary optimization technique was a financial one—namely the installation of Delni units or transceiver concentrators so as to decrease the per unit cost of each network connection. It wasn't conceivable that the bandwidth could be a limitation. With the introduction of significant applications (that is, medical imaging, distributed databases, videoconferencing, and multimedia) and the high cost of disk storage, client/server computing shifted the bottleneck to the network channel. High-powered workstations, diskless nodes, remote program booting and loading, and centralized disk storage saturated the LAN. At that time—1985—the network was still a LAN, but with up to 300 users.

The bandwidth limitations on Ethernet and the new phenomenon of saturation enhanced the reputation of the more predictable Token-Ring. Although Ethernet had a distinctly shorter transmission latency at low loads, the Token-Ring protocol guaranteed a reasonable level of performance under all conditions. Even today, this is the fundamental difference between persistent (Ethernet) and non-persistent token-passing (FDDI and Token-Ring) protocols.

Reduce the workload

An enterprise network that is saturated on the backbone, or has segments that are saturated, can see a 100 percent performance improvement by doubling the capacity of those channels. Consider the cost and effort involved in doubling the bandwidth of the infrastructure. Now consider reducing the bandwidth on those segments by 50 percent. You can double the capacity now. Be careful, though, as you do not have double the bandwidth of the original network, but you do have the capacity to double the reduced traffic levels without additional infrastructure.

You can reduce workloads by filtering traffic, changing priorities, rerouting some traffic on other existing channels, reducing what work actually is performed, installing more efficient interconnectivity devices, and replicating databases to improve not only the performance of the enterprise backbone, but also the local performance of operations using those databases. Refer to "Generalized optimization techniques" on page 195 for more information about reducing traffic levels.

One of the most important databases for enterprise networks is the resources directory. This database helps users, applications, management tasks, and anybody testing performance locate other users, applications, ongoing tasks, and databases. It is the phone book of the network. NetWare calls this the bindery or NetWare directory services (NDS), Vines calls it the StreetTalk (STDA) service or ENS, IBM calls this domain name system (DNS), while Sun Microsystems and UNIX refer to these services as network information services (NIS), formerly Yellow Pages. These tools represent the roots for universal X.400 address services and X.500 directory services. Few of the current directory services comply with these ITU standards, and it is not clear whether the standards will be useful. Nonetheless, as Chapter 4 explained, the need for information in a useful format is fundamental for performance tuning.

The resource directory maintains user account and password information, and increasingly contains information about all objects on the enterprise. When remote logins and access to services are delayed by security authentication, replicating and distribution the database reduces enterprise network traffic and reduces the time required to complete the database searches. Usually, synchronization is once per day, and this does not typically represent a substantial network load. See "Replication services" on page 259.

Shift the paradigm

Decentralize computer processing and storage. Despite the image that the enterprise network provides geographically dispersed access to anything from anywhere, this does not mean that everything important on the enterprise should be in one place. Not only is this approach irresponsible, but it enforces a model of computer centers with servers, hosts, and centralized computation, which is typically a performance bottleneck for the enterprise network. There is no bandwidth or medium that can handle a focused overload. Even Fibre Channel, with its Gbits/s of bandwidth, cannot overcome the bandwidth limitations of a host FEP, a client workstation bus, or the I/O traffic jam in a superserver.

First, disasters do occur and single-site, centralized facilities are security nightmares. Almost anything can happen—usually that which hasn't been considered possible—and it does, as Typhoon Iniki, Hurricane Andrew, the World Trade Center bombing, the floods in France, northern Italy, and the U.S. midwest, and mud slides and avalanches in Switzerland have demonstrated. Disperse operations for safety, redundancy, and to level the loads.

Second, understand that as much as technology has changed, the solutions today are much the same as forty years ago. Synchronous communications is still efficient and less likely to create the traffic jams typical of asynchronous enterprise network traffic. SNA, X Windows, and client/server to a lesser degree minimize bandwidth requirements. Centralized program and file service and diskless clients are bandwidth-intensive. Peer-to-peer computing to some degree flattens the bandwidth requirements, but it also distributes the burden for security, backup, and management.

Third, decentralization helps you design a scalable infrastructure, one that can grow and support changing technology. Decentralization provides the backdrop for supporting multivendor, heterogeneous environments. However, this dispersion with network integration does not mean you can avoid standardization with unifying protocols and architectures. Scalability provides reliability, high availability, quick recovery, built-in data protection, and the ability to replicate a service when and where needed. You might consider the performance benefits of hierarchical storage management (HSM), which can actually improve access to data and off-load some of the performance bottlenecks of single-point server and host backup schemes. The drawback of HSM is that it can trade actual network and wire performance with increased difficulty in maintaining the network. The five levels of hierarchical storage management are:

- Level 1: Primary file migration to a secondary backup device with transparent retrieval. Manages one level of the storage hierarchy (primary and secondary).
- Level 2: Disk threshold balancing; local storage management encryption with real-time, dynamic load balancing of disk space based on multiple previously defined threshold (automated storage space management). Manages *two or more* levels of the storage hierarchy (primary and secondary).
- Level 3: Threshold balancing across multilevel hierarchy; volume management; optical and tape device support. Provides for transparent management of three or more levels of the storage hierarchy, not just primary storage. Storage thresholds between different levels in the hierarchy are dynamically balanced and managed. Performs volume management including media management, job queueing, and vice performance optimization. Supports optical and tape devices.
- Level 4: Policy management; multiple platform distributed storage management. Policy management (which includes rules and file classification) and administration at all levels of the hierarchy. Provides for storage management of diverse platforms, extended from file servers to personal workstations and application servers. These services include maintaining the ownership and location of data, thus achieving local transparency.
- Level 5: Object-centric, not file-centric. Object-level management (includes structured or nonstructured records and nonfile structures). Preserves the relationships of objects at all levels in the hierarchy.

Microsegmentation

Ethernet technology predominated because it was less expensive, nonproprietary, and well-entrenched. ARCNET infiltrated the low-end PC LAN market. The primary method for optimizing LAN performance since that time has been by subdividing the overloaded network into smaller segments and installing bridges (or a PC with two NICs as a gateway) to interconnect these; this restructuring is called *microsegmentation*. Figure 5.22 illustrates the single-segment LAN with a performance overload. Figure 5.23 shows microsegmentation with a connectivity device such as a repeater, bridge, router, or gateway. You might also refer to Figures 3.18 and 3.19 for the structure of a high-speed switched LAN. At this level of enterprise network design, the bottlenecks are on the LAN segments and subnets or the intersection of the enterprise backbone and these structures.

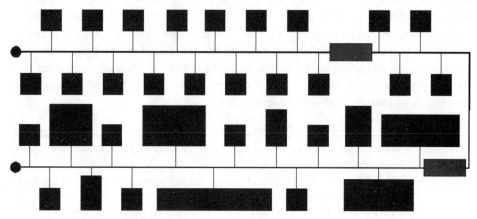

Figure 5.22 A single-segment LAN overloaded by aggregate bandwidth and excessive latency due to the two in-line repeater units.

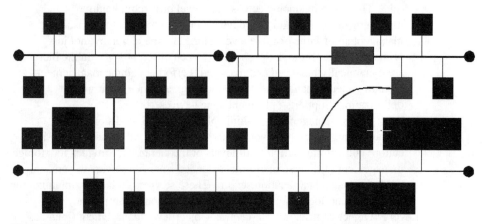

Figure 5.23 The same LAN microsegmented with repeaters, bridges, routers, and/or gateways.

Decrease the transmission latency

The real performance killer, the real bottleneck, on enterprise networks is the delay in the router or gateway queues; lack of bandwidth is rarely the killer except for WAN links or environments supporting high-bandwidth applications or the aggregation of bandwidth without appropriate network design. Rather, reduce the latency for transmission at the interconnectivity devices. This is very relevant and very effective, because most enterprise-wide communications are composed of a request, a response, a fulfillment, and an acknowledgment or confirmation; this is not a simple one-way delivery. Latency is mostly a factor of the time required to pass through an interconnectivity device. Faster devices yield a double return, because interconnectivity is two-way.

If you recall the packet transfer time and packet transmission time equations from Chapter 3, you can conclude that bridges with high forwarding rates for small packets within a well-designed enterprise network design will minimize time spent in transfer between source and destination. This provides the lowest latency. For example, a true wire-speed bridge that can handle packets ranging from 64 to 4202 bytes needs to be able to process a packet in 32 µs, but still forward the largest Token-Ring packets in 2 ms. A first approximation of a wire-speed bridge should have a single packet latency of less than 64 µs for 64-byte packets, and less than 2 ms for the 4202-byte packets.

However, at the LAN level, latency is a factor of the protocol. Shifting from 4 Mbits/s Token-Ring to 16 Mbits/s Token-Ring can save from 4 to 150 µs, basically the difference of the faster TRT. Shifting to FDDI sets the maximum TRT to 1.6 µs. However, a shift from Ethernet to Token-Ring is likely to improve performance if the Ethernet is saturated, but increase latency if the Ethernet was lightly loaded. Latency on WAN links to remote offices and distant sites can be reduced with a dedicated router connection. The link is always active, barring problems. You can improve connection time with ISDN because of its nearly instantaneous call setup, though you need to recall that latency for the packet transfer remains a function of the speed of light, regardless of the transmission protocol or the linkage. This is pertinent for on-demand bandwidth or occasional connections; see "ISDN call setup" on page 214 for further details.

Collapse the backbone

Traditional LAN technology has allowed individual departments and workgroups to deploy and maintain their own networks, as Figure 5.24 illustrates.

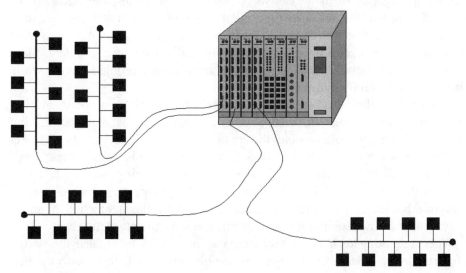

Figure 5.24 The LAN collapses into a backbone hub.

However, the issue is different when these segments are integrated into an enterprise network. While various devices exist to provide connectivity between segments (repeaters, bridges, routers, etc.), the collapsed backbone provides wiring integration, scalability, and performance enhancements. Specifically, the collapsed backbone, or *backplane*, can provide a high-speed channel for interconnecting these LANs at a bandwidth in excess of that provided by FDDI. When traffic levels are balanced on and between segments, a router (in a hub) can provide a high throughput for the enterprise network, as Figure 5.25 illustrates.

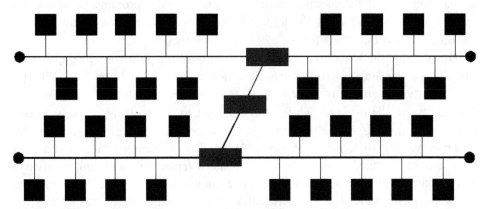

Figure 5.25 The prior single-segment LAN partitioned with multiport routers.

Increase the transmission bandwidth

A higher-speed protocol is so often the "obvious" solution. Make sure you do not fall for the myth that all bottlenecks are caused by channel bandwidth limitations. First, higher speed does not necessarily improve overall latency, nor does it improve throughput when the active constraint is the nodes, the NICs on the nodes, the design of the network, or the burden placed on the network. Furthermore, the faster protocols do not necessarily integrate well into your infrastructure, and useful interoperability is typically a maturity problem. More often than not, bandwidth limitations are channel latency problems as shown in Chapter 3. So much for that warning.

When bandwidth capacity is the bottleneck, the problem is not so much a killer bandwidth application, such as CAD, imaging, or database access, but rather the total traffic volume of all the applications. This is particularly prevalent for client/server applications and typical office workloads. When these loads cannot be reduced or shifted to off-peak, consider replacing bottlenecked segments with faster protocols. Also consider redesigning the infrastructure, as enterprise bottlenecks are more frequently caused by the interaction of many delays than the time required for the transmission of segments themselves.

When migrating to a faster protocol, you may need to replace the wiring infrastructure or modernize the legacy of the old protocol. If this seems the best course of action, recognize the cost of the transition in terms of wire, NICs, drivers, software, installation, and maintenance. A careful assessment may show that you can preserve your investment in Ethernet rather than migrating fully to FDDI or TCNS. 100BaseVG-AnyLAN also provides an integration path for Token-Ring as well as Ethernet; it is supposed to transport both protocols in the native format. Note that the 100BaseVG-AnyLAN synchronous protocol is called *demand priority*, and is different from Ethernet 802.3.

If you must recable, consider installing per EIA/TIA 568/569 recommendations for Category 5 copper or even fiber in order to provide an upgrade path for future migrations. On the other hand, if you are creating a backbone to cover extended metropolitan distances, fiber will meet that requirement, and (when used with FDDI) will also provide a more reliable and redundant linkage than Ethernet or Token-Ring on FOIRL.

Increasing bandwidth with a faster or different protocol should free up those bandwidth gridlocks. However, verify that the network infrastructure, the NICs, the hubs and routers, and other devices in the transmission path are themselves optimized for the greater bandwidth. Also, realize that greater bandwidth may not solve routing headaches and accompanying delays if the routing decisions are not distributed throughout the network globally. For example, a shift to ATM may just make it more possible to bottleneck a router with more routing decisions than it can handle, and add layers of processing to the enterprise. While ATM switches can switch ATM cells at wire speed, routing decisions are generally at a network or data protocol level; cells will need to be assembled into real packets, and these need not be decoded for routing information. Additionally, ATM virtual networking are represented by several different configurations. Specifically, vendors have produced ATM connections that provide:

- LAN switching
- LAN emulation
- Edge routing
- Virtual routing
- Relational cloud connectivity

These different options provide vastly different performance effects on your enterprise. While all these ATM switch designs provide for switched connectivity and premise wiring consolidation, only the routing options yield low latencies between LANs, an important concern for this book. The routing options and the relation cloud allow you to configure virtual subnetworks, another important consideration if you must balance your LAN loads and filter this traffic from overloaded backbones and WAN connections.

If the software, configuration, and vendor release is not fully compatible, you may not see the immediate performance increases. For example, upgrade the routers with more buffer memory so that they can accept the faster stream of packets and not drop them. Lost packets eat up more bandwidth and increase latency disproportionately. The same is true for a server or workstation; you want to make sure it can keep pace with the traffic arrivals.[3]

Bandwidth in and of itself may not be the problem. Overloaded servers, routers, and gateways cannot use more linkage bandwidth if they themselves are overloaded. Confirm that the bottleneck is a bandwidth overload. The shared media technology may induce the bandwidth bottlenecks. Refer to "Micro-segmentation" on page 220 for the tried-and-tested option for circumventing bandwidth bottlenecks. Also full-duplex, switching, and better use of inter-connectivity devices can improve throughput as the next sections prescribe.

One issue you should research very carefully is the actual availability for a particular service from your primary telecommunications vendor, because it is usually cheaper and easier to maintain a single relationship than multiple ones for WAN connectivity. Also, the equipment and protocols can be integrated through similar (or the same) PBX and CSU/DSU equipment. Of paramount concern is that the vendor actually has a point-of-presence (POP) for a particular service. In some metropolitan areas, you can get ISDN and SMDS, but not frame relay, SONET, or ATM. Check that there is a POP for ATM before committing your organization to this technology, and that remote sites can also connect into the same ATM switching network or any reliable and interconnected one.

A new service called isoENET (IEEE 802.9) provides Ethernet and ATM WAN services with speeds up to 6.144 Mbits/s over standard phone lines. This provides an integrated data, video, and telephony at a reasonable LAN-to-LAN connection speed. This technology is an isochronous Ethernet that does rely on the standard collision detection.

The issue of WAN connections is not only a question of available bandwidth but also the duration during the day of that bandwidth. Analog-based modem lines are virtually free (about $30/month) until you pay for connection time. At $0.18/minute, the break-even for shifting to a dedicated PVC is 36 hours of utilization. Furthermore, few modems can sustain more than 14.4 Kbits/s reliably, and the differential for a digital 56 Kbits/s PVC is cost-free. See also "ISDN call setup" on page 214 and "Financial performance considerations for WAN links" on page 274 for related topics.

Frame relay, ATM, and Fibre Channel are not the only futuristic WAN and LAN connectivity methods with high-speed bandwidth. Frame relay is not really high-speed in that it is available only at fractional T-1 speeds. ATM is still not a real product given the lack of ATM-to-LAN adaption layers. Primarily, Fibre Channel is a point-to-point local area service that provides almost infinite parallel capacity, and is not the media for long-distance linkages. It provides three classifications of service, which, as you can see, are geared to localized backplane connectivity, much like a HyperChannel or a hub switching matrix:

---

[3]*Can Servers Be Slower On FDDI than Ethernet?* Art Wittman, *Network Computing,* P. 150, November 1993.

- Class 1: circuit-switched/connection-oriented service that operates much like a dedicated link. When in use, a Class 1 path is available only to nodes that originate the connection. A typical use of a Class 1 connection is between a host and a mass storage device or from a host to a display terminal.
- Class 2: connectionless/acknowledged service that operates like a multiplexed link, with frames from many stations transmitted across the same link. Class 2 connections provide guaranteed delivery and acknowledgments. Class 2 connections are found typically in switched systems.
- Class 3: connectionless/unacknowledged service with no acknowledgments generated, and maximized for data throughput efficiency. Class 3 connections are used typically when data is broadcast or multicast from one node to many.

However, a concept called *dark fiber* is a nonamplified medium with an intrinsic capacity of at least 25,000 Gbits/s. It bypasses the inherent bandwidth and latency limitations of fiber networks, with electronic amplifiers and repeaters that are limited to about 2.5 Gbits/s and required every 22 miles (for SONET). It is called dark fiber because it essentially functions like a free-wheeling Ethernet (i.e., with limited command and control information encased in the messages). Figure 5.26 illustrates the ideal application and performance ranges for different transmission protocols.

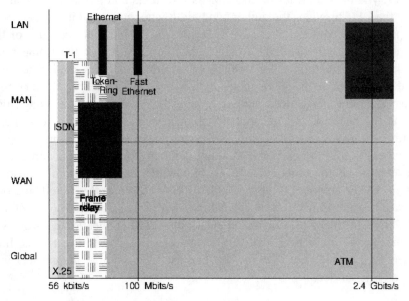

Figure 5.26 Ideal application for common network transmission protocols.

Bandwidth on demand

Having flexible bandwidth would solve many LAN and enterprise network bottlenecks. While it is not directly possible to allocate increased local connection bandwidth for LANs even with ATM (because of the committed information rate and the congestion algorithms), bandwidth can be increased with either analog or digital dial-up connections. Modem or PVC and SVC lines can be automatically activated to handle overloads. Routers can be configured to activate standby links or spanning trees when conditions require this.

While dial-up connections with modems are one possibility, ISDN provides a better and less expensive option. Not only can it be used for WAN links, but you can install for local overflow as well. It is less expensive than a leased line with comparable bandwidth. The basic ISDN service is BRI, but it is expandable to PRI. Furthermore, because it is a switched connection, you do not pay connection charges when it is not active. The ISDN call setup is 250 ms for local connections but always less than 3 s. ISDN also provides two-way connections with a third channel generally reserved for all connection overhead.

Frame relay is the ultimate bandwidth-on-demand option. Connections can be established at a minimal committed information rate, while bursts can be transported by the bandwidth allocation, *if available*. This means that if you install a frame relay servie for 56 Kbits/s with an upper bandwidth allocation of 256 Kbits/s, you have a buffer of 200 Kbits/s for bursts and peaks. . . if you are lucky and the LEC provides the higher bandwidth when you need it.

The question of congestion control is serious for frame relay because of the limited number of tools available to frame handlers. Although the frame relay protocol has been streamlined for throughput, minimum latency, and efficiency, the handlers are not robust enough to control adjacent nodes or flow control mechanisms. Frame relay does support congestion avoidance and congestion recovery at the network and transport levels to prevent the onset of overloads, and to prevent network cascade failure in the case of severe congestion.

One caveat is that the carrier has committed only the minimum throughput and does not have to provide the peak bandwidth at any particular instant. Similarly, ATM is supposed to be available (during 1994) on this tiered structure for bandwidth from 45 Mbits/s to 155 Mbits/s with a possibility of 622 Mbits/s and 1.2 Gbits/s for bursts. It remains unclear how ATM congestion controls will manage the bursts, multicasts, switched virtual circuits, and traffic floods, how the carriers will provide the overflow beyond the CIR, or how the carriers will actually sell this service.

NICs

As Chapter 3 related, NICs can create a server or gateway bottleneck when the CPU overhead to support the traffic transmission (and reception) loads exceeds available resources and forces the delay of basic data processing and disk I/O. Increased traffic levels possible with multiple NICs or tuned drivers may also create a second-order effect of a CPU bottleneck; more traffic means higher CPU requirements to actually compute and access the disk. This is prevalent when a server or gateway has multiple NICs or 32-bit NICs optimized for throughput. These units may have increased the overall traffic throughput, but at a cost of increased processing power. This creates a *subsequent* bottleneck at the CPU. Bus mastering NICs do not necessarily relieve this problem; they may increase the CPU requirements strictly for traffic transmission, or increase the CPU load related to processing the data. These are two different problems, one NIC-related, and one which due to CPU underpowerment.

Solutions NIC-caused CPU overloads is to select (compatible) NICs with the lowest load on the CPU. When the increased traffic flow demands additional CPU resources for processing and disk activity, the solution is to increase the server horsepower with a more powerful CPU, or shift to SMP-based servers. You might also consider splitting loads between multiple CPUs.

Repeater
Repeaters are designed to extend a weak signal for longer distances. They are a vestige of the need to financially optimize the early networks. Although repeaters are faster than bridges, routers, and gateways for regenerating signals, they should not be used to connect segments where performance bottlenecks are evident for two reasons. They increase latency and bandwidth utilization. Repeaters increase latency by extending the interval during which Ethernet collisions can permissibly occur, and add more devices to rings, thereby increasing the token rotation time. Furthermore, repeaters create traffic loads equal to the sum of the loads on all connected segments, and therefore increase the bandwidth utilization.

Gateway
Technically, an ISO gateway is a device that translates applications and protocols between disparate environments, such a hardware and/or software device installed to connect a LAN to a host. That is a common and important facility. (By the way, gateways according to the ISO model also provide services at all levels of the ISO seven-layer model.) Traditionally, gateways have filtered traffic and minimized cross-segment traffic when the detrimental effects were first realized. Gateways are usually standard network stations or file servers configured with two NICs, as shown in Figure 5.27. This technique is effective, although expensive, because another node must be added at the junction of the two networks. This juncture is usually inconvenient because it might be a hallway, an office, or a closet, although more lately wiring consolidation and security concerns have made it possible to place gateways in centralized computer centers.

Gateways create management overhead problems and long latencies. Maintaining /ETC/HOSTS tables, NIS, and device naming services with user changes, adds, and deletions rapidly becomes unwieldy with multiple gateways. When gateways provide primary functionality as a file server and secondary utility as a communication switch, the double CPU and I/O is better served by a stand-alone solution, such as the bridge.

Bridge

The bridge is essentially a repeater that forwards packets without a local address, as shown in Figure 5.28. It was initially designed to correct the phase jitter problem on transmission signals between Token-Ring subnets. However, the filtering algorithm provides an improvement over the repeater in that it does not create the aggregate network load over all segments.

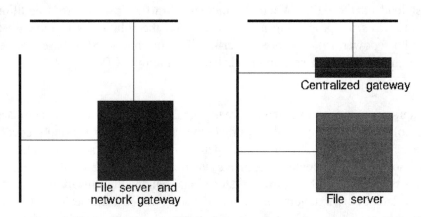

Figure 5.27 A gateway filters traffic by destination address.

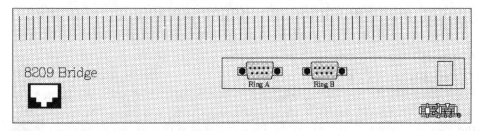

Figure 5.28 A bridge improves on the repeater by forwarding only those packets with nonlocal address. This improves network bandwidth relative to the repeater at costs in terms of longer regenerative latencies.

You improve bridging performance by maintaining scrupulous network address tables (if they are not self-addressing), simple configuration, and keeping network changes to a minimum. The bridge solves the microsegment performance problem where the media and protocol is the same. However, because this book is about internets, the issue is also about integrating diverse environments. The bridge cannot connect Ethernet to Token-Ring, nor can it handle native SNA traffic unless the bridge is specifically designed for Token-Ring. Furthermore, the bridge cannot handle addresses internal to the data packet or different NOS protocols. It just passes them along, and along, and along. . . .

Router

In 1987, the new ISO model raised the possibility of a router as a more effective means than bridges to connect segments, and less expensive than more-generalized gateways. The router repeats packets based on destination address, and also by priority, security, and administrative configuration. It can convert frame packet formats, and when designed as a wiring concentrator, it also provides simplification of multiple-segment wiring, as illustrated in Figure 5.29.

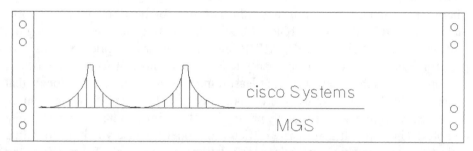

Figure 5.29 A router improves on the bridge by matching packet address to those in its routing table. It decreases network bandwidth, but because it is slower than a bridge, increases transmission latency and may drop (lose) data.

This niche market exploded to fill the demand to segregate workgroups and link them together for resource-sharing. Vendors created filtering bridges, smart bridges, trouters, brouters, filtering bridging repeaters, and other aglomerations of repeaters, bridges, and routers, gateways for the expanding LAN market. Although these hybrids still exist, the utility of the OSI model and acceptance of its naming conventions has demystified vendor marketing somewhat.

Be wary of hybrid devices such as hub/routers, routers in a backplane, and wiring concentrators with bridges. These devices are typically good for small LAN environments with WAN links, but do not usually have the horsepower for the enterprise network. If such devices are installed on the other side of firewalls, the administrative simplicity may be very useful. As soon as a hub attaches to a well-utilized T-1 line, the hybrid devices are likely to be underpowered and best replaced with single-purpose routers. You might consider the hybrid device as the perfect connection point for remote sites and branch offices, but certainly avoid attaching them to high-speed backbones.

As Chapter 3 explained, problems with router firmware, software, or chipsets cause networks to misbehave, function suboptimally (slowly), or create extraneous traffic. Before embarking on any complex or time-consuming optimization expeditions, review compatibility, integration, and implementation issues. You do not want to go in the wrong direction when the problem may be as simple as upgrading (or even downgrading) the router chipsets or software for compatibility with the enterprise infrastructure. This is particularly relevant if you are using the newer network protocols, such as FDDI, frame relay, or ATM, which are neither stable nor fully interoperable in a multivendor environment.

Just two years ago, the consensus was that routers were sufficient for the infrastructure of the enterprise network.[4] Router technology resolves the bandwidth issue for that older generation of clients and workstations. However, in compensation it created new management, reliability, integration, and latency problems. Furthermore, more powerful NICs, CPUs, and bandwidth-demanding applications require more bandwidth than hierarchical subnetting and better workload-to-latency ratios no longer solved by the fastest bridge or router.

Routing algorithms

There are a number of routing algorithms in use for host and LAN networks including IP, IPX, RIP, EGP, IGRP, OSPF, edge routing, source route bridging, DECnet, IS-IS, NLSP, RTMP, AURP, and other more proprietary ones. For performance reasons you will probably reduce the number of routing algorithms supported across the enterprise, or at least minimize the number of protocols that reach the backbone. Firewalls are good dividers between different routing protocols. Selecting which routing protocol is the primary one is actually quite complex. They are extraordinarily difficult to model because you have to define their behavior mathematically in a simulation environment. There are many trade-offs and performance ramifications, as well as security, reliability, and availability questions. You might refer to a specific source for router protocols.[5]

Watch out for non-routable protocols. The transport protocols NetBEUI and UDP are not routable and create all sorts of network havoc. UDP (which runs as part of TCP/IP) is connectionless, and as such can easily saturate segments (cannot be routed. . . remember?) with empty network calories. UDP is the primary SNMP alert and data delivery mechanism. You may want to track UDP packet loading and banish UDP from your enterprise segments. NetBEUI is the underlying LAN Manager (and NTAS) protocol, but this does not mean that you cannot *substitute* IP or IPX instead.

SNA is the native IBM routing scheme based on session names. Although widely used, it is convoluted and complex because it supports load balancing, congestion control, prioritization, and parallel connections. It does not support dynamic routing, and will create time-outs when routed or tunneled with LAN-type traffic. It is best to run it in parallel (either side-band or out-of-band) with native LAN protocols, rather than tunneling LAN protocols into SNA or SNA into LAN protocols. If you can move from SNA and the related APPN and LU-naming conventions to TCP/IP or even an IPX/SPX environment, you will have an easier-to-manage enterprise and a more uniform environment for optimization. Although SNA with prioritization and congestion control could benefit LAN traffic, the encapsulation process and overhead is extensive.

---

[4]*Ethernet Bridges and Routers*, Scott Bradner, *Data Communications*, Pp.58–70, February 1992.
[5]*Routing in Today's Internetworks,* Mark Dickie, Van Nostrand Reinhold, N.Y., 1994.

Similarly, encapsulation of SNA into TCP/IP fails to provide the link congestion control or communication with FEP devices until the wrapper is stripped away.

Boundary routing is the preferred method for attaching remote sites because it eliminates the need for the remote sites to download, update, and maintain routing tables. See "Optimizing hub and spoke networks" on page 274. The distinction that makes a boundary routing device a router rather than a bridge is that it talks to a router at the central site. The limitation of this device is that it only provides for one point-to-point connection.

Another good remote connectivity protocol is source route bridging (SRB). Some people want to pigeonhole this protocol as a bridging protocol rather than a routing one. It really doesn't matter too much, except that it does communicate with routers. Source route bridging is very good for hub and spoke connectivity because it simplifies the management of remote sites. It is simple and proven, although it does not support load balancing or dynamic routing, and does not adapt to spanning trees. Avoid source route bridging across enterprise backbones, because it uses explorer packets in a broadcast search, which can easily represent between 20 and 30 percent of network traffic. On a complex enterprise network, the explorer packet can consume between 50 and 80 percent of network bandwidth seeking the optimum delivery route. Although, you may typically see such high utilization only when nodes first insert into the network, inadequate configuration also creates these same loads.

Avoid source route bridging when the enterprise has a moderate-to-high base-level traffic, because source route bridging explorer packets easily represent between 20 and 30 percent of network traffic with the "all segment" broadcasts. On a complex enterprise network in extreme circumstances, the explorer packet can consume between 50 and 80 percent of network bandwidth seeking the optimum delivery route when devices first login, and this decreases to about 10 to 20 percent depending upon network reliability, stability, and configuration. Each new device basically says "hello" to every intermediate node on the *extended* network. It is certainly not the protocol for remote, mobile, or telecommuting connections. In a typical IBM and LAN shop, the explorer packet will consume about 10 percent of bandwidth. By the way, if you run SNA over NetWare on the enterprise network, each SNA explorer packet does a name address query for all located subnets. If you actually have spanning trees or redundant paths (active or not), the explorer packets are multiplied on all paths and returns because the RIF does not turn off the broadcast. Because of this, one explorer packet can become several hundred, all of which must be acknowledged and eventually processed to extract the optimal route. Most routers support multiple protocols and maintain multiple routing tables. Selection of which routing protocols to include in your enterprise is briefly discussed in Chapter 5; however, for the most part, environment, applications, integration requirements, and equipment determine the routing protocols.

For the most part, the SNMP protocol performance depends upon how fast the device agents can handle requests. Service time is really not dependent on the size of the request so that a query with a complex argument is handled just as quickly as each of many simple requests. But, it is still a hardware function. If you are using SNMP or SNMP version 2, you can optimize channel bandwidth utilization and device overhead. Although many proponents of this management protocol will assert that the bandwidth requirement for SNMP messages are trivial, they gloss over the effects of cumulative effects from many requests. One approach to optimize bandwidth is to use the GetBulk operation as this compiles multiple requests and replies into one message and response. Party abstraction (Manager-to-Manager MIB, RFC 1451) solves MIB sharing limitations.

First path routing is more efficient than the SRB, as are manual assignments. One advantage of manual routing assignments—that is, building the routing tables by hand—is that there is limited network overhead. Note that automatic routing table construction for some Wellfleet routers on complex networks has literally taken days to initially build or reconstruct after the network panic. The tradeoff is that a person must set up and maintain all the routing tables; formats vary from device to device as do the relative directions to the next router.

IP encapsulation is effective due to its pseudo-standardization and tenure, support for load balancing, dynamic routing, and spanning trees. It can support priority transmissions, criteria-based routing, and hot standby router activation and rerouting. Note, however, that the IP packet has undergone a redesign for 32-bit address support, and the IETF is seeking rapid deployment of this enhancement. IP encapsulation and many routing devices may not be backward compatible with the format change. Factor this into your future enterprise plans.

Microsegmentation devices

The following table summarizes the performance characteristics of micro-segmentation devices. As you increase the complexity of the segment connection, the configuration, integration, and management overhead increases, so that the software gateway is much more complex than the hardware-based repeater. There are a few warnings, cautions, and fine print about interpreting the chart. The values do not specifically reflect your environment.

Microsegmentation device	Affect on bandwidth	Affect on latency
Repeater	Increases aggregate	Slight increase
Bridge	Decreases aggregate	Doubles latency
Router	Decreases aggregate	Triples latency
Switch	Increases aggregate	Extremely variable
Gateway	Decreases aggregate	> Triples latency

Specifically, bridges and routers can create broadcast storms in excess of bandwidth reduction when routing tables are updated or a network panic starts. The gateway can reduce bandwidth requirements by intelligently filtering traffic with algorithms customized to the specific environment, by tunneling protocols, by protocol spoofing, and by quenching loads at sources as network performance degradation begins.

Hub

A hub is a wiring concentrator. Primarily, this device provides wiring convenience and integration of LAN rings or buses into a single wiring closet and wiring device. Sometimes hubs sold by vendors are chassis with backplanes providing 100 Mbits/s or more. You can get different services depending on the peripheral plug-in cards you install. Protocols supported vary by product and vendor, but usually include a subset of Ethernet, 10/100Base-T, 100BaseVG-AnyLAN, full-duplex Ethernet, Token-Ring, and FDDI.

Primarily, a hub provides microsegmentation of LANs into 2, 3, or more segments. This is called *port switching*. An integrated bridging, routing, or switching module provides the actual interconnectivity between segments. Vendors include Kalpana, 3Com, Madge, Alantec, Lannet, SMC, and Cabletron. Generally, you can increase the bandwidth of a single LAN with the built-in microsegmentation, and the throughput achieved with any of the connectivity modules tops out at between 16 and 17 Mbits/s. This throughput varies little based on whether the hub connects the segments with bridging, routing, and switching.

Vendors are providing better management tools for hubs. Because they represent an integrated and single-source product, the software management tools are beginning to show a good representation of the hub and all its modules. SNMP and remote monitoring MIBs makes it possible to actually see the alerts and lights on the panels. This is certainly very useful and appealing when the enterprise spans a wide geographical area, and does help debug bottlenecks and test possible optimization techniques. When hubs include microsegmentation or node switching, the visual management makes it possible to immediately see the effectiveness of load balancing as you do it.

Some network managers may be uncomfortable with putting so many functions in a single box. Although some hubs include redundant backplanes, switching power supplies, and the ability to hot-swap a failed module, consider the risk of concentrating repeater, bridge, router, and switching functions.

It is possible that multiple NICs and full-duplex switching could bypass the network I/O bottleneck. The available benchmarks for this configuration should be 25 Mbits/s, but were based on PERFORM3; this block load transfer does not compare well to real-world workload characterizations. Nonetheless, this might be a useful method to improve the performance of a LAN segment attached to the enterprise network. However, throughput is limited by the designs and the external limitations of connecting multiple segments into a higher-speed backbone without adequate buffering or compensation for the speed differential.

If you have been using Synoptics, Cabletron, or Networth hubs as the central building blocks to your premise LANs, you have been benefiting from microsegmentation. In fact, one of the reasons these concentrators work so well as LANs grow larger is that they compel you to structure the wiring, create hierarchical subnets, and microsegment. This is good. However, as your LANs become integrated into other and larger networks to form the organization enterprise network, the limitations of microsegmentation will become more apparent as the internal bridging or routing technology in the hubs is overwhelmed. When this happens, understand the inherent metaphor—you have the structured wiring already in place. Simply reorganize your LANs and the interconnections, offload the overloaded hubs, and install high-capacity routers, switches, or gateways in place of the more primitive hubs.

Buffered network I/O

Enterprise network traffic is different from highway traffic. Although cars and trucks in gridlock stay there and remain a part of the gridlock, network traffic disappears for a time (before reappearing as retransmissions) because it is dropped and sessions are lost when there is a transit delay. One of the problems of enterprise network gridlock is not that such bottlenecks spontaneously vanish as soon as they occur, but rather the retransmissions and gridlock management adds to the mess in the form of the network panic. The best solution in cases like this is to buffer the packets in transit, as the model in Chapter 4 detailed the effectiveness of traffic buffering.

The buffers are typically part of routers and high-end switching hubs. In effect, the buffers accept the flood of traffic much like a reservoir and release it so as not to overwhelm the destination paths. Recall the illustration in Chapter 2 that compared pressure with throughput. Figure 5.30 illustrates the elastic effects of buffers on traffic pressures and why you might consider their relevant use for performance enhancements in the enterprise design. The buffer and its size converts excessive pressure into a buffered volume without creating output overflow, and has a tremendous increase on performance by reducing the number of dropped packets.

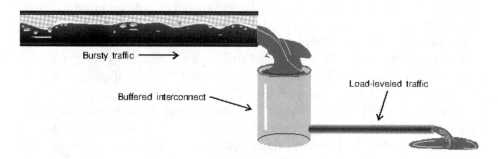

Bursty traffic ⟶

Buffered interconnect

Load-leveled traffic

Figure 5.30 A buffer relieves the pressure and damage from a flood of traffic and provides a smooth transition between segments and subnets.

Predictive pipelining and NIC enhancements

Cogent Data Technologies provides an Ethernet NIC that supports a simple collision-avoidance algorithm. The algorithm tracks collisions and dropped packets and make an assessment about how much load (and when that load) can be sustained. This extends the TCP enhancement to the Ethernet protocol. See "steady-state algorithms" on page 236. Because software represents 90 percent of the time involved in I/O (disk, network, and graphics), you might view this enhancement as a set of better packet drivers that tries to optimize this significant critical path.

Similar enhancements for NICs are available from SMC and 3Com, which provide "simul-tasking" and parallel tasking on bus-mastering adapter cards. These reduce the amount of time required to packetize and decode the data by processing before the incoming traffic or outgoing message is fully received by the NIC. Additionally, these and other vendors provide a dual-ported buffer for simultaneous reads and writes. The performance increase in terms of bandwidth is not substantial because of inherent I/O bottlenecks, but latency is reduced.

When the NIC is the bottleneck, consider adding multiple cards. The cards need not even be on different microsegmented subnets; the cards could all address the same channel. Although you might see some improvement with bus-mastering NICs, more than likely you will shift a CPU overload to a bus contention traffic jam. However, multiple cards (with different network addresses) can increase enterprise network I/O with a decreasing utility for each added card. Balance NLM for NetWare makes it possible to load-balance the NICs so that the overall throughput is more uniform and less chaotic. Note that load balancing does not refer to level workloads on a machine, but rather the workloads piped through multiple NICs.

One of the surprising effects of bus mastering NICs or other NIC enhancements is the change in server or workloading CPU loads. *Some* of these devices increase channel throughput while decreasing the overhead on the host CPU. In effect, you may increase channel bandwidth utilization, which may have seemed to be the original limitation in your enterprise network, when actually you have decreased the load on CPUs and provided them the ability to do more computing or process more disk traffic, and thus, secondarily, have increased channel utilization. Make sure you understand the bottleneck.

A new service called isoENET (IEEE 802.9) provides Ethernet and ATM WAN services with speeds up to 6.144 Mbits/s over standard phone lines. This provides an integrated data, video, and telephony at a reasonable LAN-to-LAN connection speed. This technology is an isochronous Ethernet, which does rely on the standard collision detection.

Virtual switching

Shared media LAN technology can be reinvigorated with matrix switching technologies, as Chapter 2 illustrated. If bandwidth is fully utilized, a switch can dedicate the full protocol bandwidth to many pairs of communicating nodes simultaneously. The switch creates a virtual circuit between a single source and single destination for the time required to complete a transmission. This is called *segment switching* (not to be confused with port switching, which is a type of microsegmentation within a hub or backplane). In other words, these two communicating nodes have been removed from the shared media network where other nodes are competing for bandwidth, and placed on their own private LAN for a temporary period (as long as necessary to deliver a frame), as Figure 5.31 illustrates. This "dedicated" LAN is virtual in that it exists only as long as necessary to deliver the frame. The switch improves performance by increasing aggregate throughput and connection latency, decreasing the error or collision rate, and also typically decreasing the lost packet count.

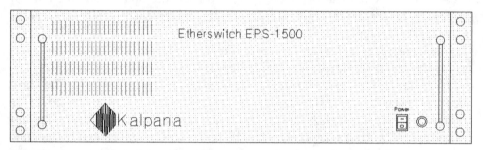

Figure 5.31 A switch creates a private LAN for communicating nodes for the duration of a frame transmission.

It is important to differentiate switching from routing, and to distinguish those products that actually switch packets in hardware at wire speed with minimal latency. Packet switching occurs at the OSI media level, whereas routing takes place at the network and data levels. Typically, switching occurs in hardware with RISC (or ASIC) chips. Currently, there are three different MAC-layer Ethernet switching options. I assume that when switching products are available for Token-Ring and other protocols, they, too, will use the same options. The current switching options include:

- Cut-through mode
- Modified cut-through mode
- Store-and-forward mode

The cut-through mode has the least latency because the MAC address is striped from the header on-the-fly, and redirected to the destination at wire-speed. It provides no store-and-forward buffering at all, and is fastest at low traffic levels. When traffic is high, very bursty, or has a high error rate, the switch forwards bad packets, and drops all those it cannot immediately forward.

The modified cut-through mode reads the entire packet header so that it can retry a failed transmission more than once. Packets that arrive for the same destination during this interval are dropped, but the buffered packet is delivered.

The store-and-forward method—also a MAC-layer bridge—has a substantial delay because arriving packets are fully buffered and checked for errors before they are switched, and when switched they vie for slot time on either the destination node or destination subnet. Note that packets can be switched not only between paired nodes, but also between paired subnets.

The primary benefit of the cut-through technology is that the switching is handled in hardware at wire-speed. While this is true for Ethernet switching now, it will also be true when switches are built for Token-Ring, FDDI, Fast Ethernet, and ATM. One primary limitation to recognize with switching (and also bridging and routing) is that it can only switch packets between network nodes or network channels if the destination channel has slot capacity at the very instant the switch has the packet, as is shown in Figure 5.32.

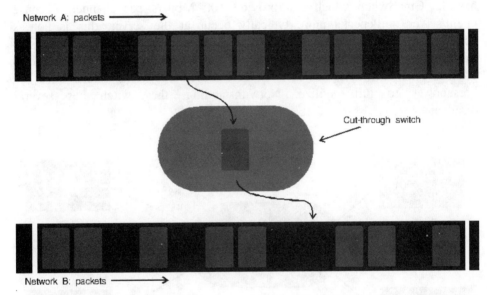

Figure 5.32 The cut-through switch is very fast because it only reads the address at the beginning of the packet. However, it requires slot time or node access to complete the transmission successfully.

Although the buffered switch may not seem so attractive because of its "excessive" latencies, you need to realize that it has different trade-offs. It does handle broadcasts (unlike the cut-through and modified cut-through switches) because it can read the entire packet, buffer it during a time interval, and then establish multiple sequential virtual connections to emulate the broadcast. It also filters unnecessary overhead and misrouted packets to reduce the net enterprise traffic load. Vendors may not clearly differentiate their switching products, and so allow you to believe the unspoken assumption that the fastest switch is the best. This is not always the case (as the queueing models in Chapter 4 demonstrate) because you often need the buffer—as explained in "Buffered network I/O" on page 222—for a transitional induction between nodes and segments to level the loads, and a means to resolve broadcasts or multicasts. By the way, you can always create a buffered switch configuration by sandwiching each unit between high-end routers, as Figure 5.33 illustrates.

Although the cut-through switch is technically faster, there are four significant benefits from the store-and-forward switch. First, the store-and-forward switch can handle the bursty traffic from the source by buffering it. Second, the store-and-forward switch can handle the bursty traffic at the destination by buffering the outgoing packets too. Third, the switch filters incomplete or defective packets by throwing them away. Fourth, the buffered switch can provide for broadcasts with sequential virtual connections. This action is shown in Figure 5.34.

Bandwidth-hungry peer-to-peer applications benefit most from this technology. The switch (which is usually part of a hub) is inherently scalable. DEC has the GigaSwitch which can provide 100 Mbits/s per channel. Because client/server bottlenecks more typically occur at the server, this is not as appropriate. Also, very few network devices can create more than 4 Mbits/s of traffic; you will want to assess the sensitivity of the bandwidth traffic jam before providing a bandwidth that few servers, routers, or workstations have the horsepower to fully utilize. Notwithstanding, the switch has several disadvantages.

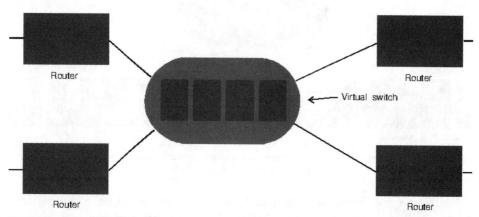

Figure 5.33 You can bypass the network latency when you have multiple routers with a centrally connected switch rather than a subnet, or emulate the performance of a buffer switch with surrounding routers.

Overall, it provides more bandwidth than the basic shared media environment, but at the expense of increased packet latency. Packet latency in currently available switches range from 0.3 μs to 770 μs, depending on whether the technology is cut-through switching or buffered store-and-forward switching. An independent test lab measured these Ethernet switch latency values: LANNET MultiNet is 13 μs, Atlantec PowerHub is 120 μs, while Triticom SwitchIt was 475 μs, all transporting 457-byte packets with PERFORM3. Note that packet loss (dropped Ethernet packets) ranged up to 67 percent of the traffic and utilization reached 65 percent of Ethernet 10Base-T bandwidth.

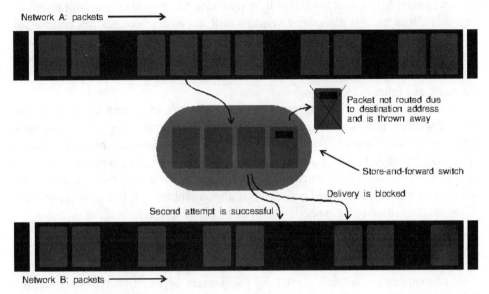

Figure 5.34 The store-and-forward switch handles destination congestion in two ways; with input and output buffering, and with destination filtering.

When this technology is part of a port switching hub, network management and segment isolation is also simplified. In any event, before committing to a switching infrastructure, test the switch performance with your network workloads, designs, and routes. Performance typically does vary from the optimal benchmarks provided by the manufacturers.

The switch provides another unique service on the enterprise network; it allows you to attach a network analyzer to any segment at virtually any time. This means you do not need to have multiple analyzers, cart one from location to location, or worry about the firewall effects of bridges, routers, and gateways. You can just switch the analyzer into the segment you want to view. However, there is a downside to this ability, as described in Chapter 3. Since each switched virtual circuit consists of only two nodes at a time, it is absurdly difficult to capture protocol information with a protocol analysis tool *between* two nodes. You cannot eavesdrop or listen in to the conversation without building a subnet (with a bridge, router, or gateway). You are limited, given the current crop of monitoring software and hardware, to the native protocol monitoring tools built into the switch. On the other hand, RMON or other vendor-specific management MIBs may provide some *limited* VLAN monitoring capability.

Furthermore, as both Chapter 2 and Chapter 3 depicted, the switch blocks out other connections during its duration. Of equal importance, monitoring the performance of a switched connection is no simple matter. Although the protocol analyzer is very effective for a shared media environment, it is flexible and inexpensive enough to attach to every possible connection. In all likelihood, device agents, MIBs, and SNMP will be extended for virtual connection monitoring. Also, it is not yet clear how vendors will implement congestion control for broadcast messages on switching infrastructures. If a single device is the bottleneck, even if that bottleneck is on the wire, switched connections are more apt to increase the magnitude of the bottleneck so long as traffic is centrally sourced or directed, as Figure 5.35 shows.

Realize that all the benefits of virtual switching are lost when traffic is routed over a shared media segment. At that point, all the limitations and contentions inherent in Ethernet or Token-Ring apply to traffic routed between virtual segments. Since it is likely that preexisting shared media segments and subnets are part of the enterprise network, you might try to avoid this limitation by keeping these segments and subnets on the periphery of the core enterprise network. The shared media sections can be terminal subnets.

By the way, the actual hardware implementations for packet switching are designed around shared-memory systems, shared-bus designs, or multistage matrix switches. Shared memory and shared-bus buffer I/O provides a common pool of memory for actually switching the packets between ports. However, the shared-bus design usually incorporates a bus bandwidth significantly higher than the total bandwidth required for the aggregate bandwidth of all ports. This is useful for handling internal bursts. The multistage matrix switch is an array of switches, each supporting two inputs, two outputs, and an in-line (but out-of-band) control lines. The switches are usually geared to ATM and are particularly effective because they can be designed to the fixed-sized data payload. And, they are fast, simple, and lower in cost. It usually does not provide the throughput under load that shared-memory or shared-bus designs can, but on the other hand, it can handle overlapping traffic in the matrix fabric without blocking. Although support, protocol implementation, and other factors may predicate the selection of one vendor over another, nonetheless, understand the implications.

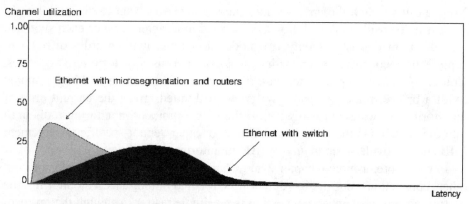

Figure 5.35 A switch can decrease performance when the traffic is unevenly distributed.

Be careful with switches when designing a bypass of bottlenecks. You may inadvertently create some other ones. Although switches connect multiple networks, the architecture is usually designed so that a switch is really a backbone with multiport connections. Because a switch is effectively a two-port bridge inside a backbone, the switch adds an extra hop to the transmission. Be sure to count the actual physical hops for the internal bridging. Additionally, many protocols, IPX among them, have the transmission window set to one acknowledgement for each transmitted packet. For those designs, a buffered switch will increase latency and degrade performance, while the cut-through switch can increase performance, subject to all the other switching concerns. The bottom line is that you will need to model, simulate, or test switch performance within your unique enterprise network architecture.

Port switching

Dedicated matrix switching establishes a temporary virtual network. It is an OSI level 2 function. On the other hand, this same term has been applied to wiring hubs that allow the network manager to manually (via a toggle switch) determine to which segment each port is functionally attached. Port switching is also called *dynamic switching*. Some are even configurable remotely. A port or *segment switch* provides LAN microsegmentation. It does not increase the bandwidth. It does not dedicate bandwidth for virtual connections. It does not create a temporary circuit. It is still a shared media LAN technology. It does increase the available bandwidth available to a network by increasing the number of physical segments or subnets that exist. So to that degree, the port switch does increase the available bandwidth per device.

Many large hubs, for example the Synoptics 5000 or the Cabletron MMAC, provide port switching. It provides the same opportunity for optimizing enterprise performance as microsegmentation. It is, however, simpler to toggle some switches than to pull patch cables, figure out where they should go, and thus create logical subnets or rings. It is more reliable, more configurable, more flexible, and certainly more scalable.

Furthermore, port switching is useful for establishing virtual LAN configurations. With dynamic work groups, team-oriented development projects, and transient organization structures, devices, services, and workstations can be dynamically redistributed to different segments. This facility simplifies network configuration and management, but is still a microsegmentation technique.

Full-duplex transmission

Full-duplex transmission means that dedicated inbound and outbound channels are established for a bidirectional transmission at the protocol bandwidth. Duplex transmission is currently available for Ethernet and for Token-Ring. Kalpana is the primary vendor for duplexed Ethernet, while IBM is the single supplier for LANStreamer full-duplex Token-Ring. It is a technology that should scale well for any protocol and transmission speed combination. In other words, duplex

Ethernet means that two directional 10 Mbits/s connections are established between devices, as shown in Figure 5.36.

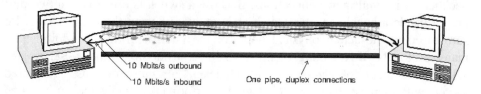

10 Mbits/s outbound

10 Mbits/s inbound

One pipe, duplex connections

Figure 5.36 Duplex connections decrease Ethernet latency to yield a significant performance increase for peer-level devices communicating with each other.

This technology is useful for connecting busy enterprise routers and hub or parallel servers that communicate with each other. The main benefit is not so much that the channel bandwidth has been doubled—remember, it is directional—but rather that the collision mechanism for Ethernet has been disabled (really, bypassed) so as to decrease transmission latency. Sustainable bidirectional transmissions at 10 Mbits/s are possible only with superscaler servers, hosts, and high-end workstations that support a multithreaded, multitasking operating system. Integration and interoperability problems may limit how useful this technology will be, although IEEE committees are trying to standardize the full-duplex protocols proposed for Ethernet and Token-Ring.

Even Pentium systems cannot sustain an intelligent transmission level greater than around 4 Mbits/s; nonetheless, this may represent 100 percent effective loading on an Ethernet or Token-Ring segment. It may only represent about 25 percent of the effective loading on a virtual network. While PERFORM2, PERFORM3, and other randomized block transfer tests can utilize about 17 Mbits/s of the available 20 Mbits/s, more realistic benchmarks cannot sustain such throughputs. Again, the major benefit of this technology is where the aggregate network traffic exceeds available bandwidth or where the unidirectional nature of most LAN protocols preclude simultaneous transmission and benefit from reduced round-trip latency.

A proposed variation on duplex technology is IEEE 802.9a, which includes the standard 10 Mbits/s Ethernet and a second separate isochronous 6 Mbits/s Ethernet channel for voice, video, or time-dependent transmissions. Isochronous nodes and standard nodes cannot communicate across the channels. While this seems an interesting concept, existing networks may be better served with other architectures or doubling existing structures.

Power bandwidth

It seems so obvious that ATM, which has at least 45 Mbits/s capacity, can relieve an overburdened Ethernet or Token-Ring. At 155 or 622 Mbits/s it can make FDDI and Fast Ethernet look lazy. However, you will need to resolve several serious design and integration decisions. Specifically, backward explicit congestion notification (BECN) and forward explicit congestion notification (FECN) are two congestion control protocols that will work together with frame relay and ATM links. BECN is the better choice, because it provides source-quench rather than FECN longer and more complicated messaging path. Virtual circuit flow control (VCFC) messaging looks very similar to a router broadcast storm; try to avoid this protocol. Variable bit rate (CBR) and variable bit rate (VBR) promise less of a delayed-feedback problem, while the committed bandwidth protocol provides less randomness. Alas, ATM cannot perform address resolution.

Optimization details

The remainder of Chapter 5 discusses nuts and bolts details for saving money, tuning the network operating system, optimizing your WAN or remote site connections, increasing application performance, and working with the limitation inherent to any enterprise network architecture. While *Computer Performance Optimization* provides extensive information for tuning workstations and servers, this chapter details how you tune client/server applications, replicated databases, and mixed traffic in the enterprise environment. Some of these details increase energy usage, cooling requirements, wiring, transaction control, software tuning, database optimization, and both client and server performance tuning.

Energy usage

Energy usage can kill the enterprise network. Most organizations will gladly pay the cost for the electricity, but underestimate the overhead for the power grid, the UPS support, and the HVAC infrastructure needed to cool the enterprise infrastructure. The CPU and monitor typically consume about 10 percent of the total workstation energy, and the numbers that you typically see for workstation energy requirements represent only a third of the total infrastructure energy requirements for that single workstation. When you add in the resources required to support intermediate nodes and monitoring equipment, energy usage is about four times larger. Figure 5.37 illustrates this pyramidal energy usage.

Power consumption for the average PC is 200 watts sustained and about 120 watts for the average monitor. Printers consume about 120 watts. The organization with 1000 nodes can save $25,500 annually with strategies of energy conservation and compliance with EPA Energy Star programs. Every watt actually consumed represents 3.3 watts of total demand when the waste heat production (and removing that waste heat through air conditioning and cooling) is factored into the energy equation; this translates into $76,500 yearly through indirect electrical savings. Overheating and power fluctuations kills network devices. In addition to creating a blueprint for the basic infrastructure, LANBuild also calculates power requirements and cooling needs, as Figure 5.38 illustrates.

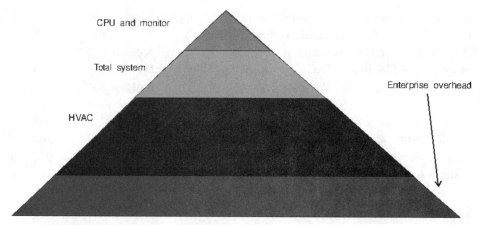

Figure 5.37 A CPU and monitor represent only 10 percent of system energy usage, while all systems represent only a third of overall infrastructure energy requirements.

Power supplies in servers, monitors, printers, and workstations convert AC current to DC. These transformers are not efficient, converting 70 percent of the load into heat and only 30 percent into electricity for the computer circuitry. Newer power supplies conform to the International Electrochemical Commission IEC 555-2 standard for matching power supplies closer to power requirements and creating more efficient switching power supplies. While the EPA Energy Star compliance program is primarily designed to reduce the need for additional power plants and lower CO_2, SO_2, and NO pollution, you are more likely to view this waste as heat that needs to be removed from your network environment.

Saving energy with the current technology is not always possible. Many CPU and keyboard idle detection programs incorrectly determine that a system is idle during computation- or communication-intensive processes. Disable the feature in order to maintain a normal performance level. Two devices solve some particular problems. The first is called 110 Alert, the other Green Keeper.

110 Alert

110 Alert does not save money, cut energy, or anything particularly active. It measures the internal temperature of any computer and beeps while the temperature exceeds 110° Fahrenheit. This is useful for connectivity devices in remote wiring closets or servers, to assure their availability. The connectivity devices can easily become unstable when they overheat and are ignored. PCs used as routers or even standalone routers that overheat create network panic as the routing tables and protocol filters suffer from the senility caused by RAM overheating. I used several of these devices to check the performance effects of loading up client workstations and servers with adapters, and found that most chassis are not designed for multiple NICs or other adapters disrupting airflow. Cards with daughter boards or lots of RAM are even more disruptive. Figure 5.39

shows the 110 Alert; you might notice the standard Molex power connectors for installation in most PC, NuBus, or even proprietary systems.

A LANBuild Design and Planning Tool Report
Network Installation Bill of Materials
for Ethernet network

Date: 09-10-1993 10:43:12

Unit	Description	Quantity	Notes	Spares
Network	KW electricity	216		65
Network	Dedicated circuits	302		91
Network	Tons BTU cooling capacity	192		68

Figure 5.38 LANBuild calculates energy and cooling requirements.

I like these simple devices because they have prevented a few problems from becoming crises—and saved me. It takes a while to figure out what is buzzing when one of these alerts sounds off because they are otherwise silent and invisible. It is not really an overheating problem, but rather an air circulation problem. Move the 110 Alert around inside the computer case to silence it, or explore whether the computer device really is at risk for overheating. Some Energy Star computers cycle down the CPU and lower the power supply and fan speed. As a result, the internal temperatures heat up beyond 110° Fahrenheit.

Figure 5.39 The 110 Alert checks for internal operating temperatures that exceed 110° Fahrenheit.

Green Keeper

This device plugs between the keyboard and monitors keyboard activity. When the system is idle it switches off the monitor and accessory devices (such as a printer or external modem). There are competitors to the Green Keeper product, but they work the same way. Payback for the $50 device is about three years. It is important to keep in perspective that electricity is only one component. The cooling requirements are reduced as well. In all fairness, it is better to manually switch off the monitor and printer and turn off the idled CPU at night, or when it's not in use for more than an hour. That will save $75 in three years instead.

However, this is a good concept for servers and connectivity device monitors, which are typically monochrome and are frequently on 24 hours per day. Green Keeper will sense the idle keyboard and turn the monitor off. This will preserve the low-end amber or green devices from burn-in associated with static displays.

Optimizing desktops

WIN.INI and SYSTEM.INI can be optimized to improve MS Windows performance marginally. The most substantial change is to shorten them (by removing extraneous entries) to speed up loading time. Files that exceed 64K exceed limitation and may not load completely. Most users win by minimizing their use of fonts. Not only does this improve overall desktop performance, there tends to be similar results with any programs that access fonts (such as Word, WordPerfect, Excel, etc.). Many commands within these files are added when you add new software or change drivers; therefore, you want to make changes to these files through Control Panel functions, or special .INI editors or tools that track adds, changes, and deletions to these critical files. Several examples are included on disk and referenced in *Computer Performance Optimization*.

Optimizing WANs

Model your traffic levels, as described in Chapter 4. Filter traffic and create a firewall between segments on the enterprise network. You need to model your traffic load, packet sizes, and latencies to see whether the bridge, router, or switch will provide the better performance. The longer latency of the router than either the bridge or switch, while material, may be more than offset by the better management of traffic levels, but it all depends upon your network parameters.

If the WAN transmission channel is limited in capacity, the most potent strategy is to reduce line overhead and increase payload delivery. Typically, you want to minimize RIP, SAP, and NLSP services with filtering, but also reduce the number of concurrently supported protocols. By the way, RIP/SAP filtering is built into the NetWate Multiprotol Router SAPCFG.NLM. Additionally, consider the NLSP or IP replacements for the tried-and-true IPX services. You increase payload delivery with LIP, data compression, packet header compression, and packet burst transmission utilities. As stated previously, you do not want to use both software data compression and data compression built into the DSU/CSU or router hardware as this will increase the transmission latency and potentially bottleneck the connectivity devices.

Use NetWare link services protocol for WAN routing rather than RIP/SAP; it lessens intersegment traffic load and cuts down on the overhead traditionally associated with verbose routers. This protocol provides for the establishment of a routing domain and minimizes traffic broadcast enterprise-wide for routing table updates. The tables and network layout are more stable. Secondarily, this technology should contain the effects of a network panic. If you use the NetWare burst mode protocol, optimize the setting for the burst mode window. However, note that the BNETX drivers are deemed flawed by Novell and no longer supported; you will want to use the PBURST VLMs. This configuration value can be increased or decreased dynamically. However, remember that this protocol requires an extra 102 bytes in protocol overhead and additional RAM on servers and clients. The benefits of stream data transmission outweigh the extra overhead only when the majority of the multihop packets are greater than 128 bytes. The burst mode NCP is sensitive to buffer counts and sizes. Too few buffers and you slow performance, while too many buffers and you drop packets. Hence, use the dynamic settings to optimize this protocol, because it is easy to reduce enterprise network performance with this protocol extension.

Connecting LANs

Although you must tell most servers and network devices that they are actually on different network segments or subnets by giving each segment or subnet a unique address, you also want to make certain that all devices on each segment or subnet share the same segment or subnet address. Even though different addressing causes a significant performance hit for the LAN and the servers alone, this problem is more pronounced when bridges, routers, switches, and gateways bounce packets from misaddressed machines around the enterprise. Normalizing segment addresses when in error will reduce bandwidth utilization and reduce latency for some transmissions in a very big way. If some pairs of nodes seem to exhibit particularly bad performance, you might look at the segment addresses.

Optimizing with multiprotocol stacks

The best method to improve enterprise performance is to unify the protocols in use and select one global protocol. UNIX, VMS, Vines, and ES9000 support TCP/IP for interconnectivity and interoperability, and TCP/IP is usually the most uniform and widely-applied protocol. If using this exclusively is not possible, it may be possible to route and translate protocols at gateways, thus limiting traffic on any one segment to a single protocol. This represents the next best technique. Even SNA traffic can be tunneled in IPX or IP by a gateway; IBM provides this service with its AnyNet software. Some segments use AppleTalk, DECnet, or OSI (for X.500 directory services) and can also support TCP/IP as a secondary protocol. However, it is usually better to add a native service at the host so that it communicates directly, using either IPX or IP with its partners.

However, NetWare products are predominately IPX, which, although a derivative of IP, does not interoperate well. There is also the difference between NetBEUI and NetBIOS, again similar but incompatible protocols. As a result, some devices will need to support two or more protocols so that they can communicate with other devices on different segments or subnets. While loading multiple protocol stacks is usually feasible for UNIX and OS/2 with flat memory address space and the most reliable options, installing a multiprotocol stack driver is more efficient for DOS and Windows. Specifically, ODI (Novell) and NDIS (Microsoft, etc.) use less memory and run faster than concurrently loaded protocol drivers. In some cases, both ODI and NDIS can be loaded for interoperability and interconnectivity.

The use of sockets (as with WinSock) also yields substantial performance improvements by effectively bypassing the layers of the protocol and building and decoding packets directly. This excessive overhead (and robustness) is one of the reasons OSI stacks have not displaced TCP/IP, NetBEUI, NetBIOS, and other common protocols. However, the programming techniques for sockets are marginally complicated, and you will want to make sure to do not write to an unsupported or unconnected socket. You will also want to check all return codes from API calls. As multitasking, multithreaded client/server applications propagate throughout the enterprise network, you will also want to standardize port numbers, addresses, and socket addresses. Header blocks for file descriptors, calls, packet structures, and for security and authentication will protect the network from rogue programmers and keep ad-hoc efforts from competing for overlapping and duplicate resources. By the way, not all applications will need a socket interface; you can also use the readily-available RPC services in UNIX and NTAS environments.

Optimizing network operating systems

Operating systems share many similarities, and thus many of the optimization techniques and pieces of advice offered here for one environment can often be abstracted for other ones. Caching, memory management, protocol stacks, workload reduction, and design are key performance targets for all environments. The next few sections describe some common environments and methods for improving performance.

TCP/IP drivers

TCP/IP is not monolithic. Not only are there many sources for TCP/IP packet drivers, there are many different implementations of it as well. Furthermore, there is at least one fundamental change (1987) to the TCP connection and steady-state algorithms that provides a learned congestion-avoidance when buffer space becomes overrun or the network is congested. This prevents lost connections or degraded performance.

Because most TCP implementations conform to this change, more pertinent issues are the memory requirements, driver size, driver speed, and performance characteristics. There are many DOS and MS Windows 3.x stacks available; these TCP/IP drivers are important if you want to provide Internet Mosaic or Gopher access or multiprotocol connectivity and interoperability for users. Specifically, TCP/IP can load as part of the IPX/SPX packet driver binding, as a DOS application, a DOS TSR, a DOS device driver, or as a Windows device driver. Each method trades memory with speed, task switching times, ability to unload the protocol, or the ability to run multiple instances of the drivers.

Increase the TCP segment size and the TCP window size to the maximum the NIC and device can address. The TCP window size is the more important (and the more memory-intensive) parameter. This setting allows multiple transmissions with only a single response needed; it is very effective for decreasing overall background traffic in a client/server or demanding X Windows environment. Some legacy LAN cards will not support more than one segment and one window. If performance is critical, replace that NIC with a unit supporting multiplexed sessions and multiple connection streams. Performance will be most significant on high-delay networks such as the WAN and the enterprise network. Optimize NetBIOS and TCP connections for the number you anticipate (at least 7); this will optimize RAM and optimize the using of multiplex sessions over single streams.

Many routers were designed and tuned for TCP/IP over many years. As a result they provide only a fifth of the performance for routing IPX. The enterprise network may benefit from a shift to IP rather than the RIP/SAP and IPX protocols. Also, Novell uses a hybrid Ethernet V.2 and 802.3 packet format. Translating bridges often have trouble encapsulating them, and a tuning solution might include replacing bridges with routers that convert packets directly.

Replace SLIP connections with PPP. The serial connection is more robust and supports data compression, modem line speed negotiation, routing and filtering of packets, data encryption, and on-demand dial-out and call-back.

Also, you might note that many physical connection protocols, such as ISDN or ATM, are sensitive to the network protocol. NetWare running IPX will work much faster when replaced with the native TCP/IP drivers and their sliding windows in an ATM or ISDN environment. In fact, you may reach B-channel capacity with TCP/IP, in contrast to 30 to 40 percent of capacity with IPX. Furthermore, compression (for PBX, LAN, and WAN traffic) works too, and you might double the 64 Kbits/s of a B-channel in real-time, although with some increased latency required by the compression and decoding algorithms.

Multiple stacks
Running multiple stacks for IPX/SPX and TCP/IP creates several performance bottlenecks, as Chapter 3 explained. NetWare/IP improves some aspects of

channel performance by using native TCP/IP on Novell networks. The basic advantage is that it eliminates the flood of SAP and RIP broadcasts seen on native NetWare. When used in conjunction with burst mode and large internet packets (LIP) protocols, this is particularly beneficial, limitations of BMP and LIP notwithstanding. In addition, NetWare/IP adds the usual RCP, FTP, and other typical TCP/IP transport services. However, the extra overhead of NLM and client driver software slows server and client CPU by about 10 percent.

LAN Server users will benefit from OS/2 LAN Server 3.0x or greater because this network operating system has enterprise network support features. The primary performance improvement is based on its native support for the Intel Pentium CPU. However, the release includes support for a domain hierarchy, which is actually important in minimizing traffic between domains. For example, a master database (which supports replication) contains network layout, routing information, passwords, access rights, locations, and identities. The master Yellow Pages increases complexity of network management, but decreases the level of background traffic. The mechanism replaces the domain and backup domain configuration with its imposition of broadcast requests for user authentication and domain access to all attached domains, regardless of whether the user exists on those domains.

Large internet packets

Originally, NetWare sent packets to different networks with a maximum size of 512 bytes. These spoon-sized bites of data add overhead, increase bandwidth, and increase latency. Data sent in small packets is inefficient, particularly across SVCs or PVCs. By using the large internet packet protocol option or third-party products that aggregate small blocks into larger ones (and even compress or multiplex them), you can improve transmission efficiency.

Handicapped wait

NetWare processes (threads) can be halted or delayed with a handicapped wait. Such waits are either temporary or permanent, and are typically a feature of production code with imbedded enabled debugger commands. In some cases, these wait states are established so that NLMs do not steal all the available CPU time, but yield time to other processes. It is possible to alter the process control block or set other handicaps for improved performance.[6]

Cache configuration

Throughput results from cache efficiency and the ability of the cache to anticipate read and write requests.[7] This is not only a function of size, but also

---

[6]*Temporary and Permanent Handicap of Threads,* Novell Professional Develop, *Bullets,* March 1994.
[7]*NetWare Performance Tuning and Optimization,* Ron Lee, *NetWare Application Notes,* Pp. 1–12, October 1993.

the configuration of the caching parameters. Essentially, the tradeoff with cache is between size, the hit and miss ratio, and the locality (time-accessibility) of the information in the cache. In effect, realize that the data you want to fetch must be in a format that matches the optimal unit within the cache. If the information will not fit, you do not want to attempt to cache it. However, sooner or later, the information in the cache turns over and no longer matches the current processes. This cache hit and locality is addressed by cache conflict and associativity, the methods for locating information in the cache. Although this information is of most importance to designers of cache hardware and software, the concepts are pertinent to how you configure your cache since you are going to stress that cache hardware and software more than any other components but system CPUs.

Although NetWare reads one byte from a disk as quickly as it reads an entire block from disk, the overhead for caching blocks is higher within the cache in terms of RAM actually used and the opportunity lost to cache other data. Hence, you do want applications and operations that minimize or compress data in cache. However, there are five NetWare parameters you can optimize for better cache and disk performance. Consider changing them individually—not all at once. The same holds true for other types of network operating software or system software. Load the NetWare Monitor and review memory and resource statistics before and after making these substantial changes. Key statistics are cache hit rate, percent of RAM allocated to cache, and the NLMs actually loaded. Do not forget to unload the Monitor when no longer needed—it requires resources, of course.

You should disable the read after write verification (On/Off, On is default) if this operation is built into the controller or consider replacing existing controllers with units with RAID functionality within hardware, which is faster. The dirty disk cache delay time (0.1 to 10 s, default is 3.3 s) sets the efficiency of the elevator disk writing mechanism. Shorter delays increase disk I/O to the detriment of overall CPU performance, but longer times compromise data integrity and can also decrease performance by blocking access to disk I/O for extended periods while changed data is flushed to the disk.

Maximum current disk cache writes (10 to 100, default is 50) sets the number of cache changes that can be written to disk in a single elevator disk operation. Higher numbers improve overall server performance but to the detriment of read operations. File servers (dishing out applications or user files) will provide better service with higher values, while database servers that typically have higher read-to-write ratios perform better with lower values.

Minimum file cache buffers (20 to 1000, default is 20) specifies the number of caches specifically allocated to server operations, and subsequently the amount of RAM allocated between cache and all other uses. The amount of memory allocated to cache should be about 60 percent of total server RAM. Process-intensive server operations (SQL, for example) may require more RAM for table extraction and manipulation, whereas file-intensive server operations typically require a higher allocation for caching. The buffer values can have a positive and negative effect, and can substantially improve performance.

Turbo file allocation tables reuse wait time (0.3 s to 1:05:54 s, default 5:29 s) improves access to large or fragmented files by indexing the subsequent blocks and retaining this information in cache for configurable periods of time. Increasing the setting requires more CPU, RAM, and caching resources, but this will improve performance where large files are more frequently accessed, as with graphic or imaging operations. On the other hand, the overhead is substantial where database records are the predominant I/O, and basically irrelevant.

Cache configuration and optimization is not solely a province of NetWare. Vines also supports various file system caches including a file system cache, individual volume caches, and CPU caching. In fact, most network operating systems have parameters that represent a subset and/or superset of the NetWare caching parameters. Note, only five NetWare caching parameters were listed here; there are, in fact, about 23 parameters. Complex caching parameters is the norm for native UNIX, VMS, CICS, CMS systems and for third-party caches as well. Even workstations—including simple DOS—have several native caching configuration options, while third-party caches provide as many as 15.

Windows NT and NTAS uses virtual memory much like the VMS or UNIX operating systems. As a result, most of RAM may represent application paging or swapping of disk activity into and from the virtual memory space. When physical RAM is exhausted or overcommitted, performance degrades dramatically of course. You can best address this thrashing by increasing the disk space available for paging files or add RAM. As Chapter 4 explained, monitor the swapped pages/s counter. Sustained high values indicate thrashing. This may also indicate improperly designed server processes, databases, or application code.

Windows for Workgroups and Microsoft Office applications use OLE 2.0 functions and common "snippet" applications. For example, Visio is available both as a full application or as a pop-up component software that integrates into Word or Excel. This extensible software snippet saves disk space. However, these "applets" and .DLLs are typically non-entrant and may be configured only when Windows is initially loaded or when the .DLL is initially loaded. New instances are not loaded. Although this saves RAM and stack space, it can lead to unsatisfactory performance or inability to perform some tasks at all. As these applets become mainstream, beware the effects of applets accessed by many different applications. The attachment of applets or DLLs often means that a database engine (such as Access) cannot be reinitialized. When accessing foreign tables or external databases, this initialization must occur upfront, once, and not need to change during the duration of the session. Beware multiple initialization files (.INI) in the network environment because you may not know which one is the active and overriding configuration file.

Specialty caches (such as CD-Blitz, SpeedDisk+, or even SmartDrive) provide negligible CD-ROM improvement for client stations. Since a cache stores only the most recently accessed data, and CD-ROM access is typically sequential and random, the effectiveness is dubious at best. Write caching is not pertinent since the CD-ROM is *read only*. Frequent access to a directory is substantially improved, but that represents a rare scenario. Also, CD-ROM access on client stations has no impact on the network. Few network CD-ROM caches are available; generally, you would want to map the CD-ROM as a standard network storage device, like any other disk volume, for the best results. However, even network access to CD-ROM is still typically sequential and random.

Tracking the disconnected device

As Chapter 3 stated, there is a disproportionate effect from a disconnected device that may even precipitate the dreaded network cascade failure. While the best performance is reasonably assured by fault-tolerant or redundant systems, it is also advisable to establish alarms and performance thresholds so as to track key linkage devices, service providers, and perhaps even off-line workstations. After all, there is a physical and financial limit to providing useful spanning trees, alternate routes, and backup domain services. This is critical concern as more information processing is distributed to remote sites. Although this does not directly improve performance, at least it provides a means to automatically monitor enterprise operations and inform management and users of an impending failure. You may also want to track performance of mobile devices as a separate threshold; someday these roving units may well represent an aggregate traffic load in excess of the load levels now experienced on most LANs and enterprises.

Optimizing application code

Performance in the enterprise network environment really begins with application *code*. Too many managers, consultants, designers, and peripheral network people believe that any network performance problem can be resolved with faster protocols and transmission methods. This is usually untrue, and will not address the problem of the internetwork, as channel bandwidth may represent only one bottleneck. While FDDI will save *milliseconds* over Ethernet on the front-to-back time for a packet, this does not address the *seconds* consumed by router queues, or *minutes* of processing devoured by inefficient SQL statements and bad code. Recall the queueing models from Chapter 4 that compared the performance of a very fast switching hub with a high-end router, or the comparison of the bridge versus the router. The functional bottleneck turned out to be the front-end processor. If you take that concept a step further, you really want to look at the applications you run over the network, and the actual code itself. Do not rise above the problem; it often begins at the code. Review the code (and particularly the error trapping) to ensure that it will function with WAN links and the glitches that occur with external services. Refer to my *Visual Basic Optimization* (1995) book for additional information. A typical glitch is a user request for data that clogs the network. In those cases, a *query governor* can minimize network traffic or abend excessive SQL fetch requests. Furthermore, you want to address performance problems at the design and planning stages, because as Figure 5.40 illustrates, the costs of addressing performance escalates after deployment.

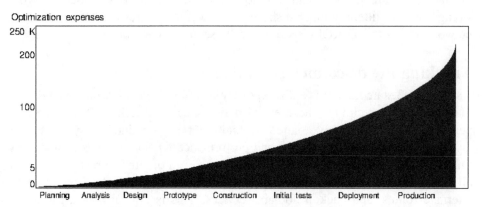

Figure 5.40 Performance optimization costs increase during the life-cycle of systems and applications.

While application code issues are beyond the scope of the network manager, they have a significant effect on network performance. Bad code creates poor enterprise performance. Moreover, code which doesn't handle errors, remote server inavailability, transmission line failures, and data replication inaccuracies or synchronization flaws represents a serious performance problem. In effect, the only means to improve latency and resolve bandwidth problems is to address the source; the source code that is.

There are various techniques for improving run-time performance of code. Usually, however, there is a tradeoff between code size, code execution speed, and the functionality of the application. There is the very obvious tradeoff made when products are pushed into production as rapidly as possible; functionality is a key design criterion on par with rapid application development (RAD) and object-oriented programming (OOP). In fact, there are quite a few new books about RAD and some large consulting organizations pushing this concept. In many ways, unfortunately, it is antithetical to performance optimization. In fact, this is why OOP development environments represent the tools of choice for client/server application development.

Some people make the strong argument that inheritance and class libraries increase the efficiency of coding by making code reusable, making it easier and faster to debug, upgrade, and maintain, and increasing reliability. Not only are the OOP tools generous with their code, execution is slowed by interrupts from states, keyboard, mouse, disk, and network. Because they are triggered by changes in object values and states, event-driven, and typically have multiple execution paths, they frequently create a functional cascade. This is not a cascade failure as occurs on the enterprise, but rather a chain reaction of code execution based on the events or triggers. Examples of popular OOP environments include Visual Basic, PowerBuilder, and C++. This concept was explored in *Computer Performance Optimization* and additional information about code bugs is explored in *Visual Basic Debugging* (1995), while *Visual Basic Optimization* (1995) focuses on methods for improving front-end VB and SQL performance and repairing cascading and suboptimal object-based code.

However, OOP doesn't have to verbose. In fact, part of the concept of C++ is that variables, functions, and definitions can be referenced and overloaded to provide multiple services. This hopefully minimizes code. Bloated code often results from designers and programmers not knowing the full set of built-in functions—and designing routines that reinvent preexisting functions—or how to apply the language commands to full effect. For example, C has a very full set of libraries for data conversion between ASCII, binary, decimal, and hex which almost makes it unnecessary to create any new routines. Similarly, most languages support logical comparisons (AND, OR, XOR, and NOT) which provide a double savings both in code volume, data heap churn, and in execution time. When you want to swap two buffers, it is not necessary to create a third temporary one. Every program swaps buffers; it is a common requirement. Rather, you can compare the two buffers with this technique:

```
A = A XOR B
B = B XOR A
A = A XOR B
```

Will not work, you say? Try it out. It works in other development environments supporting the OR or NOT OR operators, such as Visual Basic, Gupta SQL, and PowerBuilder. This is a neat trick. However, it does show the performance gains possible from almost any application. There are other tricks like these documented in programming books from the 1950s and 1960s, and recycled on Internet and CompuServe as messages or tips in the libraries.

OOP techniques as applied are not the only hazards to network performance. Applications increasingly blur the distinction between the code and the data. Objects exist on forms, in code, in data sets, in files, and in *compound* files. At the simplest level, duplicated executions of an application can no longer exist as duplicate references to a single application, because the data is part of the program and the program in effect overwrites part of itself. In other words, the application is not reentrant. This has a profound impact on OODBMS, which imbed objects and data into the code itself. A server providing client/server database operations must partition memory into an area for each client application, not just a smaller memory partition for the flag, status, data, and functional reentry points. Figure 5.41 shows how OOPs code bloats memory.

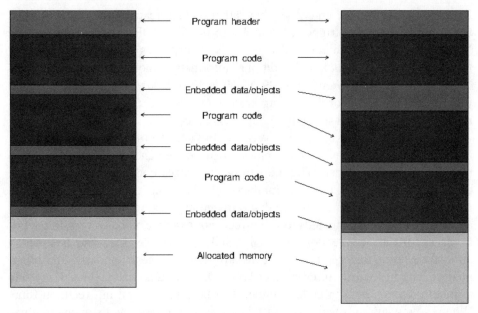

Program header

Program code

Enbedded data/objects

Program code

Enbedded data/objects

Program code

Enbedded data/objects

Allocated memory

Figure 5.41 OOP code increases server requirements because each instance of an application must reside in a separate memory partition.

However, OODBMSs typically run faster than relational designs in the TPC-A and TPC-B benchmarks. The code and the design has been optimized for speed, but at the loss of significant flexibility. While DB2 is very flexible in terms of what it can do, this is at the expense of performance. Note also that IBM provides four different DB2 products that have minimal overlap, especially on error handling, despite the common product name. See also "DB2" on page 256.

Replication of applications only represents the sewer of the enterprise network performance drain. Data created in one application can be imbedded through OLE into various other ones. More specifically, an MS Word report can include embedded spreadsheet tables, spreadsheet graphs, Visio drawings (through the Workgroup add-in functionality or through the stand-alone program), and references to the output from a complex MS Access or VB application through active OLE 2.0 linkages. Those other individual imbedded components can change during the interim and the container registration process will search the network for the status of each. As a result, you are likely to see a surge in network activity as these components are fetched and updated in the master application. Imagine the results if any of the individual components is not a master document but also an amalgam of other imbedded subcomponents. The immediate operation spawns secondary (and even tertiary) tasks unanticipated by either the user or the network designer. This results in service delays, serious ones when components are distributed widely on the enterprise.

Networks with synchronous protocols

As much as 75 percent of all SNA traffic is administrative packets. A technique called *spoofing* can reduce bandwidth requirements by having a device in the FEP or a router attached to it send back the keep-alive and administrative status messages. For example, typical SNA traffic that might normally consume 38.4 Kbits/s can consume less than 9.6 Kbits/s of bandwidth. This technique is relevant not only for remote SNA linkages dedicated to the host traffic, VSAT links, but also for remote links tunneled into TCP/IP or another LAN protocol.

A few vendors, such as Micom, Cisco, and 3Com, have SNA software and hardware protocol spoofing products to reduce the amount of traffic crossing the enterprise network. This technique also increases performance over the wide area and enterprise network by reducing the traffic latency and messaging.

If you want to integrate your protocols on the enterprise, you will probably find it more advantageous to route SNA over TCP/IP than to route TCP/IP over SNA. Not much LAN traffic can travel over the typical SNA channel. There are many ways to combine traffic. These range from time-division multiplexing, statistical multiplexing, converting SNA applications into TCP/IP-based ones, or segregating the streams within a single carrier signal. For any method, the channel(s) must have sufficient bandwidth to handle the traffic volume.

Choose the proprietary APPN over TCP/IP for distributed client/server applications in an IBM environment. APPN is not APPC or SNA. Rather, it is an IBM counterpart to TCP/IP, designed for client/server processing where a host is the server. APPN throughput tops out at wire speed and is not limited to the 0.25 Mbits/s to 2.52 Mbits/s range for many TCP and IP socket implementations. You probably need to research the APPN fit to your traffic loads and environment. In any case, APPN requires careful consideration with multiple hops. Not only does it not route as well or as quickly, but this protocol is not universally supported on enterprise connectivity devices. Non-APPN traffic (including NetBIOS and TCP/IP) can be routed over APPN if the network routers support the DLSw routing protocols, and if the DSLw supports LLC on your routers. There is some minimal evidence that DLSw will maintain SNA sessions over TCP/IP connections when the network link is saturated with other traffic.

SNA typically contains streams of text. The efficiency and latency can be improved by compressing the communication streams. This yields a 30 to 50 percent performance improvement for the end user. Additionally, if the SNA communication is reaching the limits of the WAN link, it is usually far more cost-effective to add data compression hardware and software than to pay the installation, monthly service costs, and circuit usage for more bandwidth. When LANs are backed up to a mainframe—a reasonable idea, given the precise and careful backup processes in place for mainframes—any effective bandwidth improvement and latency reduction is usually helpful so that the backup is completed within the allocated off-hours schedule. Data compression is actually very effective for SNA traffic, because the compression reduces the number of packets transmitted, and hence the number of synchronized responses.

If the enterprise is standardized on TCP/IP, then consider developing host applications to take advantage of the environment. However, if the issue is not so much integrating the different protocols, but rather consolidating the WAN linkages, consider X.25 or frame relay. You can integrate the transmission environment without merging the protocols, yet still have the option to interconnect and interoperate the host and LAN applications with a protocol that both can support. The integration would occur at the CSU/DSU, and thus the LAN and host protocols would remain separate but nonetheless multiplexed on the same channel. Although X.25 is limited to 56 Kbits/s (and somewhat less when the management overhead is factored in), frame relay can provide service as high as 256 Kbits/s, or even E-1 speeds if the LEC supports it.

One of the my favorite integration techniques for combining LAN traffic, enterprise traffic, and SNA or bisynch over WAN links is not by tunneling, spoofing, or translation. Use an older technology called multiplexing. Multiplexing the streams can be more manageable and provide better performance. Think of this mix of traffic as pedestrians, bicycles, cars, taxis, buses, delivery vehicles, and trailer trucks all vying for the same congested city street; it doesn't take too many problems to create gridlock. Instead, imagine allocating lanes to each type of traffic based on priority, burst load, and transmission speed. This gives every class of traffic its own priority lane. Hypercom and Sync provide this same facility over T-1, E-1, and fractional T-1 lines, and this is effective when the total bandwidth fits within the WAN bandwidth. Multiplexing is useful when the IBM 3725 or 317x FEP are at capacity for rerouting traffic point-to-point.

Either X.25 or frame is useful for integrating LAN and SNA traffic streams at a multiprotocol router. This bypasses the routing overheads typical with trying to concurrently transport LAN and SNA traffic. It eliminates the need for a channel multiplexer, and it can provide more sessions into the host when the FEP has reached capacity, or provide FEP-like access at a lower per connection cost than the FEP. When tunneling SNA inside IPX or TCP/IP, there are some LU 6.2 application-level options that can affect performance. Modify the transaction requirements in the applications themselves to eliminate or accelerate acknowledgments and perform concurrent conversations.[8]

Optimizing transaction processing

Transactions are the grist for many operations. If at all possible, segregate transactions streams from bursty network traffic. As Chapter 2 and Chapter 3 explained, SNA and X-Window traffic is optimized for narrow-band transmissions. Although it is managerially desirable to integrate SNA, X-Window messaging, and telephones into the enterprise infrastructure, the best performance

---

[8]SNA LU Type 6.2 Reference: Peer Protocols, IBM, Research Triangle Park, NC., 1991.

still is to separate them into their own channel. However, you might consider multiplexing or inverse multiplexing transmissions over single WAN links, and compressing the data for better utilization of the costly communication channels.

One method you do not want to do at all costs is to yo-yo between different versions of an environment, shift between compilers, or rewrite code for a seemingly faster SQL language. Albeit that there are differences between Gupta and ODBC or even other SQL dialects (and other programming languages, environments, and platforms), stay with a single course of development. Although Gupta may run faster than the PowerBuilder implementation, stay with the environment. Even when FoxPro with Rushmore or the Sybase/Microsoft SQL engine is somehow better or 30 percent faster than a competitor that you are using, the differences are not significant enough to warrant any shift. Performance is not just speed, and speed differences are quickly equalized by infrastructure baggage. What you will encounter is a disastrous shift in terms of making working code work with a new environment, retraining programmers and users, and a learning curve an order of magnitude greater than was originally invested in the processing environment.

Multithreading and multitasking code optimization

As more applications make use not only of multiple processors in servers, but also multitasking operating systems, archaic code will not run faster unless it is rebuilt to exploit the advantages and benefits of the new environments. It is important to realize that applications written for Intel 386-series processors run only marginally faster on Intel 486, Pentium, or IBM Power PC platforms unless recompiled and tuned for the new processor. Similarly, code for simpler operating and network systems will run at parity unless profiled and tuned for this environment. Refer to *Computer Performance Optimization* for more details. The concept is pertinent to C, C++, Visual Basic, UNIX, NTAS, Windows NT, Chicago, Cairo, Daytona, and OS/2. Furthermore, the move from serial to multithreaded programming requires a change in mindset. Here are some tips for improving performance of multithreaded and multitasking apps at the code level:

- Avoid global data
- Allocate data close to the code
- Avoid multiple pointers to the same data
- Avoid passing references to data between subroutines
- Use arrays and other "static" data structures
- Avoid monolithic code (large repetitive chunks)
- Keep functions short
- Avoid aliases
- Use clear loop constructs with finite ranges
- Look for parallel processing opportunities
- Understand how your threads interact and how they must synchronize

Threads that access global data run the risk of race conditions, AGI, pipeline interlocks, and deadlocking. Global data and variables also make it harder for the compiler to detect serial code that can be converted to run in parallel with other threads. Exploit static and local data structures that provide the persistence of global data with reduced visibility (OOPS encapsulation). Also, create simple relationships between data and code so that you can create parallel tracks. This makes the application easily understood, more easily debugged, and (in the long-term) more maintainable.

Although lists and trees are more elegant to program, they are harder to split into parallel processing. It's easier to fork a thousand threads than access a thousand-element array or search a thousand-node binary tree. This advice becomes important when trying to determine side effects from multitasking and multithreading or when trying to dissect large blocks of code that can run either on the client or the server, or as subtasks. Optimization may best be achieved by shifting the processing from a gridlocked server environment to the idling client.

While convenient to program, aliases mask that your code is really using shared data, with all the requisite dangers associated with local data names and structures. Avoid dependencies between separate iterations of the loop. Some-times, the straightforward serial approach hides much of the parallelism in a task that a compiler will miss. Perform unrelated operations in a different sequence to reduce data dependencies, as this provides an almost automatic compiler optimiz-ation with the dual-pipeline Pentium or multitasking and parallel-processing operating systems. Restructure your data so multiple threads can work on different parts simultaneously. If you miss one or more critical sections that require exclusive access to data, you'll create AGI-type conditions that are guaranteed to break your program. If you have too many critical sections, your code will be safe, but you'll suffer a performance penalty.

Client/server transaction processing

Treat transactions as transactions, and avoid repeating groups in the same structure. A clue to this problem is entries fields for a limited (or fixed) number of detailed atomic items, such as ORDER ITEM 1, ORDER ITEM 2, and ORDER ITEM 3, rather than a reference to related child table with a single ORDER ITEM field. If you see this repetition in the code, you can be certain the designer or programmer does not know about arrays and relational database design. This constrains the application, wastes memory, wastes disk space, and limits how the code can be upgraded. Details should be instances of a parent table. Fields for phone numbers, home numbers, main numbers, facsimile machines, modem lines, cell phones, BBSs, and the growing number of access points should represent a single repeated instance of a contact phone defined by a description and the phone number itself.

This methodology increases the integrity of the database, saves space, and makes it possible to query a database for instances without coding for each specific instance, as with IF ORDER ITEM 1 >= "toy" OR ORDER ITEM 2 > = "toy". . . . and so on. The code is smaller and faster testing a condition like IF ORDER ITEM = "toy" where ITEM.ORDER NO = ORDER.ORDER NO.

Even when data does not change often, it is advantageous to reproduce calculated values, rather than piping stored previously-calculated results from the database over the network. This saves disk space, network bandwidth and latency, but at the cost in terms of local CPU cycles (which are usually very cheap and fast). The counter-example is when the information required to make the calculations is large and slow to access, or the calculations are slow.

Tune application code

Understand the process and the inherent limitations of the environment. If you must do remote transaction processing over a 9.6 Kbits/s line, design the application to run locally and fetch only small amounts of data over the connection. Pinpoint what consumes the most time in the application and optimize that. If transmission latency for each character fetched remotely is slowing calculations, load the data in blocks instead; do not try to improve the calculation speed in this example, because it is not the bottleneck. Also, remove platform dependencies, as these are often bottlenecks again when the environment changes.

When 16-bit applications are recompiled for 32-bit environments, such as UNIX and Windows NT, you may relieve some code bottlenecks because the compilers are better, but typically applications grow in size to accommodate the 32-bit machine language. This increases loading time, RAM requirements, disk storage requirements, and sometimes execution time. This usually occurs because the application has not been optimized to the new compiler, environment, and benefits of the longer word length.

SQL

SQL is the linga franca for enterprise network and client/server applications. However, despite the ANSI standards (and there are multiple revisions of these standards), SQL performs very differentially on different platforms, across gateways, and even optimizes queries differently. You will also want to watch how a client/server query can be optimized for processing at the server, at the client, and for data transmission over the connecting wire. SQL compilers and intepreters that try to optimize queries frequently fail to improve programmer coding. In fact, the so-called optimized query may increase network demands at all components or even fail, as happens with PeopleSoft financial and human resources applications on Sybase System release 10. This may not be so much a fault of Sybase as it is a universal problem for database vendors in general.

Also, you may want to review the long queries and their composition to avoid those bad requests for data that clog the network. Again, the query governor can minimize network traffic or abend excessive SQL fetches.

The major performance problems of SQL are actually inefficiencies. For example, a query that returns many rows that are not part of the answer wastes precious server, client, and network bandwidth resources.[9] This can happen in:

- Relation scan
- Inefficient index scan
- Inefficient access path
- Extremely complex SQL query (with sub-selects)
- Non-matching index scan
- Matching index scan on non-unique index with low or poor cardinality
- Matching index scan on only part of a full key
- Wrong order join where the rows do match the initial WHERE conditions
- Junction tables
- Data tables fragmented or not clustered
- Many records marked as deleted but still in the tables

Other inefficiencies include the overhead associated with accessing the same rows multiple times. Unnecessary sorting adds to this problem. High-cost row access, as might occur with record locking, excessive paging, or security authentication, also adds overhead; this might occur when:

- Using a non-clustered index
- Table or index schedule inappropriate
- Highly skewed data input values
- Nonmatching index scan
- Duplicate key records

Typical SQL optimization techniques include alternative solutions external to SQL. For example, do joins external to DB2.2, redefine the index clustering, or create a new index (if only temporarily). You might also *remove* indexes when not needed, as each new record added must be inserted into all related indexes at a cost in terms of overhead for each index. On the other hand, add indexes on read-only systems. This is particularly pertinent for replicated databases at remote sites.

Data usage has a significant impact upon server, client, and channel usage. Consider performing data analysis. For example, you may want to review how sparse or populated the tables in your database are. Sparse tables suggest that denormalization may improve performance. Although, each access to a sparse table may represent mere milliseconds, these add up dramatically in complex queries. Here are some questions to research as part of a network performance optimization strategy:

- What tables are accessed, in what combination or order, and how frequently?
- What columns are frequently searched?
- What combinations are frequently searched?

---

[9]*DB2/2 Access Paths and SQL Tuning,* Graeme Birchall, 1994, CompuServe 73540,1566.

- What data is not used?
- How detailed (summarized, averaged, or totalled) is the data?
- How often is table or column referenced or accessed?
- What is the time frame for data reference or access (days, months, years)?

The answers determine how you define data storage, or the tables are normalized, if joins are saved and reused rather than being rebuilt time after time, and whether you can cope with some of the data anomalies in both normalized and denormalized data access. Also, you may discover that some indexes are better than others, which will have a profound effect on overall network performance. You may also discover that the views and definitions are incorrect, or at leat suboptimal to the end-user requirements. As such, you can redefine data definitions to create new data views for more efficient data access. Unused data is costly in terms of access performance, disk space, and backup. Purge unused data, or relegate it to offline or CD-ROM storage. Similarly, the time frame is important as it suggests ways to archive unneeded tables and sets, (remotely) distribute or centralize databases, or determine the priorities for HSM. Similarly, you will want to review how queries are used.

- Who wants the information?
- How often are queries run and who is running them?
- What queries are run?
- How frequently are the queries run?
- What are user expectations for response time?
- How are query results used?

Most applications reference static (optimized and compiled) queries. Ad hoc queries demand significant bandwidth in real-time, and users often repeat and fine-tune requests until they get the required results. Such bad queries waste resources and one approach you might consider is training users how to define a better query. You also might consider better front-end tools for database access; this can improve the hit rate for query-building. When you look at time requirements and time frames for queries, you might discover that some could be automated to batch jobs, controlled times, or performed at off-peak hours. Additionally, you might discover that some complex queries could be codified as a stored view or redefined for more efficient resource utilization. In the same way, you may find users misusing the output or generating information that already exists in the desired format. They are just performing duplicate work and wasting network resources. Retraining, documentation, and access to views optimizes resources.

Increase the buffer pool available for joins. One very interesting statistic is that inner joins, even though they return fewer rows and involve fewer records, are not actually any faster than outer joins; the difference is barely measurable.[10] Also, benchmarks should exclude the first tests in the series, as these are atypical of the

---

[10]*TPC Benchmark: A Full Disclosure Report*, IBM Corporation, Austin, TX, 1993.

steady-state query environment. Place conflicting data sets, log files, input files, and program code on separate disk volumes.

When using Btrieve (version 6.10 or later) with large variable length records, there is a benefit to using real-time compression from both the server and requester side. This minimizes channel bandwidth, speeds up data delivery, and lightens the load on the server, because the decompression occurs at the requester workstation.

Similarly, you can use multiple cursors to improve the overall performance of UPDATE transactions. Although complex statements are often optimized by the SQL interpreter or best optimized by scripting and saving repetitive statements, some complex transactions with unique keys can be broken into smaller steps within a BEGIN LOOP/END LOOP structure. This pipes results from the ongoing steps into subsequent ones without creating intermediate tables.

Exclusive access to databases disables record locking and other multiuser functions that merely add overhead and decrease performance. If the databases are truly exclusive, move the data from the network to a local hard disk. Local disk latency is several orders of magnitude less than the hop across the network. Do not over-index a table. Too many indexes slow addition, deletion, and updates to records as the indexes (in separate files) must be updated to reflect the changes as well. Read-only access to tables also bypasses many of the security, record locking, and multiuser features of SQL.

Expand encoded data and save it in the display format. The data lookup or table join required to match up "12" for "Radiology: CAT" is more expensive in terms of processing and network overhead than maintaining data in the useful and human presentation. Views provide better performance when they do not contain aggregate functions; that forces a recalculation when records change, and a complete review of records in the table matching indicated condition.

Increase the page time-out, buffer sizes, data read ahead, and number of retries for locked records to maximum file access and reliability. Although the host or server may have a cache in place, the SQL engine usually creates its own cache for buffering recently read or written records. Also, how each query is translated for execution is an interesting consideration. Some tools precompile each query each time at execution; this is suitable and necessary for ad hoc database searches, but typically impractical and slow for client/server applications. It is usually better if production SQL statements are tested in an interactive environment but compiled for actual networked operation. You might check the performance of these so-called optimized queries to see how they actually perform.

Another consideration is where the queried data (rows) is cached or sent. Usually each request is maintained as a temporary table on the server. This is advantageous for most client/server operations, where the results are reviewed and updated rather quickly. However, when the table is actually combined as an ad-hoc view, or manipulated and joined with other data, it is more efficient of network channel resources to FETCH the rows to the local workstation. Note that performance can be dramatically improved when volatile data in the view is specifically not updated. If it is necessary to refresh the views with updates, changes, and deletions, you will probably need to balance channel load with server performance, and see how the user must respond to the time delays and need for database consistency. By the way, this concern is very relevant when data is replicated over many database servers and the latest information is not updated locally on a frequent basis.

The number of ad hoc queries has a significant effect on database server and transmission channel resources. In an environment with 1000 users, with 25 percent performing just one ad hoc query daily, where each request actually generates 8 table queries, you can see a total of 2000 queries over the enterprise. Now consider that SQL is quite complex even with a friendly user interface, and more than likely users will need to rephrase a query five or six times to get the results as initially anticipated. Even good programmers may refine a particular query twenty times more to tune the syntax and the format within a code block. This results in a load of 16,000 to 40,000 queries daily, a profound load for any network, regardless of how great its bandwidth or how minimal its latency.

Visual Basic

Visual Basic (VB) is an amalgam of sequential code and object-oriented programming. The sequential aspect of VB is not much different from other dialects of basic, or COBOL for that matter. Performance is equivalent to most pseudo-compiled languages (including COBOL). When integrated with ODBC or SQL tools for accessing databases, it performs the same as COBOL. The basic tools are not specifically focused to financial processing, however, they provide the same results. Furthermore, the OOP aspect of VB provides real-time control features and triggers. The ability to have a real-time trigger provides asynchronous processing capability, which is very useful for talking to a host, controlling multiple tasks, or optimizing certain tasks before others. Specifically, an operation does not have to wait for a prior one to complete before initiating or performing other ones. Although MS Windows 3.x is not multitasking, VB is.

VB SQL performance can be dramatically improved by creating transaction blocks. In other words, for transactions involving a repetitive disk record creation or update, use START TRANSACTION and END TRANSACTION commands to bracket the activity. Explicit COMMIT TRANSACTION also speeds up write operations. The reason is that record changes are retained in a memory buffer until flushed to disk. Repetitive record writes implicitly flush prior ones to disks. However, when records are constructed as part of a larger series of transactions, the writes are flushed to disk only at the explicit conclusion of the entire transaction block.

One application writing about a hundred records took four minutes to complete this task; when the two lines of code were added, execution was improved to 23 seconds (1100 percent improvement). That very dramatic performance improvement is not to be expected for every application. When the transaction buffer exceeded available RAM, the creation of the temporary file provided only a 50 percent performance boost. When this trick was tried in other dialects of SQL (even Access Visual Basic), performance results varied greatly.

Note that this is a very specific enhancement that should not be confused with other different but similar-sounding terminology. For example, if you are running Visual Basic as a client of a NetWare server, you will find better performance if you SET NCP FILE COMMIT = OFF from the server console. This has nothing to do with transactions, but everything to do with fine-tuning server operations by eliminating an extra process acknowledgment and network message. By the way, this also works for Clipper and other databases too. Note that there is a minor increase in risk of incompleted database updates during a server crash; this may prevent applications from accessing the database until it has been repaired, rolled back, or backed up from backup storage media. See "COMMIT=OFF" on page 271 for other NetWare client/server fine-tuning.

Unload all add-in managers you do not want. This will greatly reduce memory consumption and create applications that do not need them either. Reduce the number of objects you use, particularly graphical ones and maintain at least two sets of images in memory for each object. Every object increases the need for system resources and slows down processing, as state values must be checked for changes. Also, do not enable an object or activate it for drag-and-drop unless you intend to use those features. This saves overhead and improves performance.

Locate the VB dead code, fat, unreferenced code, and duplicated routines. Find the rewrites of system calls that bloat programs in any language, not just Visual Basic. The not-written-here-syndrome is not only a pride issue but also one of ignorance of the available functions, APIs, and system resources within the development environment. One of the more interesting tools to pare down an overly large application is to apply VB Compress from Whippleware, because it will mark and consolidate bloated code, as Chapter 4 explained.

There are other tricks that are pertinent for database processing in VB. For example, do not use the MoveLast record command, as this is a sequential access. Attach external tables to a database rather than joining external information to a native database. It will be faster. Also, the linkage is established once and not reestablished each time the external information is referenced.

The use of assembly-code libraries or VBX extensions also provide substantial performance improvements when judiciously applied. By the way, the VBX libraries can also be used with PowerBuilder or C. Wang/Open provides the means to take scanned text at about 300 Mbytes per page and convert it with OCR into text files with 6000 characters. While OCR in general is slow, the 98 percent reduction in object size yields a tremendous savings in terms of reduced network traffic and file storage requirements.

Also, because imaging represents the next waves of increased enterprise network traffic, other methods of improving graphical performance are relevant as well. MediaKnife buffers multiple images so that VB will not reload picture and image objects repeatedly, or try to repaint them over and over. The reloading is usually network traffic-intensive and a good VBX and coding technique can reduce the traffic. Repainting is a client process, and as such should not affect the enterprise. However, better coding techniques and explicit control of repainting and redisplay will improve performance as the user experiences it. If you employ video and text wipes and transitions with MediaKnife, you often can use the time to asynchronously perform some intensive calculations or fetch data from the network server while the user is entertained; performance is often a perception as much as a technical issue.

One last trick presented here is that Visual Basic can be used as a design, rapid application development, and testing environment. All resources and forms are easily converted into C or C++ templates, and most code and functions have equivalents in Windows APIs or standard libraries. This means that a slow VB application—slow in manipulating Windows objects, processing or calculating, or I/O-bound—should run about 10 times faster in a compiled language. Refer to my *Visual Basic Debugging* (1995) and *Visual Basic Optimization* (1995) books for additional information.

PowerBuilder

PowerBuilder is not too different from other GUI, SQL, and interpreted at run-time code (P-code) applications. Many of the fundamentals presented for Visual Basic will improve PowerBuilder as well. However, PowerBuilder is more limiting in its access to databases and how it accesses tables. For example, completely normalized databases eliminate wasted space and duplicated information, but at a cost in terms of slow and complex PowerBuilder applications. Multiple table joins using complex SQL trade disk space with performance. I/O is the most limiting bottleneck. Complex screens and tables also increase the need for Windows resources. Try this:

- Design a database that minimizes the need for table joins
- Limit screen updates to only one database table at a time
- Limit database table updates to only one screen at a time

Place all attributes for data relationships in one table. PowerBuilder's "update" function will use a data window to handle everything with less overhead. After all, interpreted SQL is not as efficient as native functions. The trade-off is that the database is no longer normalized. A normalized database has common attributes in one parent table, and specific attributes in children tables, linked by foreign keys. Because PowerBuilder is OOPs code and event-driven, check your paths through the code and objects. You must make certain that updates/inserts are done to the parent table first, after which, you can do the child updates. When deleting a parent entry, the children should be completed first to avoid foreign key violations. The alternative is to construct a transaction and rollback if incomplete. This has the advantage in single- and multiuser environments of improving overall performance so long as the necessary tables are exclusively accessed, or records are not locked by other users.

Limit screens to no more then 40 objects, to minimize the P-code overhead for PowerBuilder. This includes: buttons, windows, lists, fields, and scroll bars. PowerBuilder must maintain all of the events and attributes associated with each object. This overhead consumes Windows system resources (limited to two 64K buffers) and slows the application. Too many active windows potentially lead to Windows system crashes or GDI overruns. Similarly, favor window inheritance over object inheritance. Grouping several objects onto one generic window and then inheriting from the window reduces the overhead required for Power-Builder.

Use libraries for your objects. Limit groups to an absolute maximum of 40 objects, because PowerBuilder has difficulty when there are more then 40 objects in a library. Maintain consistent naming conventions so that you can use multiple libraries and avoid creating subsets or supersets of preexisting ones.

DB2

Although this book primarily is concerned with the performance of enterprise network activities, I would be remiss not to include host processing requirements from typical large-scale databases, such as DB2. The best way to improve DB2 performance is to reduce the workload by offloading or removing redundant or immaterial tasks. The second option is to shift the processor platform to the SMP or parallel I/O models and share data through the ESCON architecture. The Starburst research project has put considerable effort into optimizing DB2/2 and DB2/6000, since these are such key migration platforms for IBM. Performance is primarily a factor of workload, caching, and a trade-off between memory, storage, speed, and complexity. Performance is also sensitive to the availability of free cache in the memory pools: however, increasing available memory for caching is a financial tradeoff, and also a technical one when you reach the maximum support memory for the host or processor architecture. Refer to Chapter 4 for a discussion about DB2 performance measurement tools for other database environments. DB2 performance tuning consists of:

- Reducing virtual storage requirements
- Reducing actual (as opposed to virtual) I/O
- Reducing machine instructions
- Reducing I/O and overhead instructions
- Increasing the data pool (cache) sizes

Precompile the SQL statements once the applications have been tested, and check out how DASD storage improves overall data throughput; because so much emphasis has been placed on throughput in the past, the IBM-proprietary storage has been optimized and tuned for efficient transaction retrieval and storage. Use EXPLAIN, an SQL preprocessor and compiler analyzer, to explore the data access plan for complex queries. Maintain active, takeover, and hot standby systems to increase reliability and integrity of the database. Employ scalable processors so that CPU overloads can be parcelled out to additional inexpensive processors. Install an intermediate device between the host and workstations for client services. This off-loads the host, as Figure 5.42 illustrates.

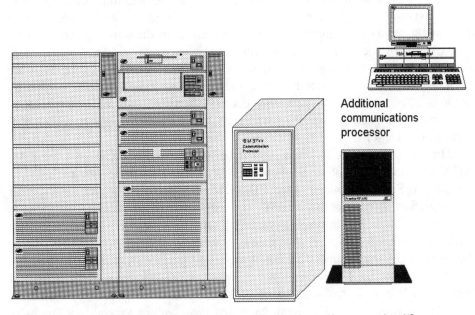

Figure 5.42 A device providing communication for the host can increase host performance and net I/O.

DB2 over SNA with the LU 6.2 peer protocol can be optimized by reprogramming to limit messages and the windows for controls, as shown at "LU 6.2 application-level options" on page 246. Typically, upgrades of significant proportions solve some specific operational or reliability problem. Some upgrades address performance issues as well. For example, the newer versions of DB2 support several buffer pool enhancements for larger central storage and more memory allocated to buffer pools. The more of DB2 that can be contained in memory, the better the performance. The newer releases provide dynamic reconfiguration of hyperpools (with the ALTER command) so that virtual storage does not need to be stopped and restarted. This saves time, but also provides the ability to monitor and tweak performance subject to database loading conditions. Also, multiple buffer pools can exist rather than just a single one so that a pool does not have to be shared among several different applications or many tables.

SNA gateway performance optimization
IBM Labs states that you need to model the network objectively, while at the same time repeating that it "is dangerous to extrapolate from these test measurements" and models. They question the validity of comparing latency

across the network with response times as experienced by the users. Nonetheless, latency is measurable, predictable, and inherently a proper statistic for enterprise network performance measurement and prediction. So use it. (The other is bandwidth as a percent of available channel bandwidth.) It is essential, according to IBM, that you take full advantage of the gateway settings—both at the FEP or other host communication processor and at the LAN gateway—so as to provide the best possible performance. However, IBM also cautions the casual manager from deviating from the preset ("optimum") default values, as it could degrade host and gateway throughput. With caution thrown to the wind, general tips for improving are included in the following few paragraphs. As always, these configuration parameters are obscure unless you have seen them before. They are included here and interpreted so that the reader will: be aware that gateway settings exist; understand that the gateway settings can have a profound performance impact, for better or worse; realize that the settings have complex tuning ramifications.

The local or remote gateway should be a dedicated node when it provides SNA connectivity to other network workstations. This node can be a PC, a specialized Synoptics adapter (see CD-ROM slide show) in a hub, or a similar configuration. A multipurpose server is feasible where latency and workload is low, but consider the options for creating a separate gateway when performance suffers. Create separate additional gateways when interactive node counts exceed 50, as the maximum supportable count per IBM OS/2 gateway. Also, realize that frame sizes on 16 Mbits/s can be as large as 16,453 bytes (although IBM technicians curiously reiterate that the MAXTSL frame that can be transmitted over Token-Ring is 15,732 bytes). Note that these values exceed 8192 bytes as the maximum inbound and outbound PIU frame size—the default buffer setting for most Token-Ring NICs, and the 4202-byte NetWare frame size. It is very important that all workstations are able to buffer the full receiving frame size to prevent more congestion than improper microsegmentation of the client workstations ever will.

By the way, when supporting remote nodes, the maximum SDLC I-field size is 521 bytes for line speeds greater than 9.6 Kbits/s. To complicate this, SDLC I-frame for Token-Ring is 2057 bytes. The NCP buffer size must be 240 bytes, whereas the RU size ranges from 1792 to 2048 bytes depending on the actual Token-Ring NICs in use. For flow control, set the SDLC window size to the maximum value (MAXOUT=7). Set virtual route pacing as VRPWSmn=(10,48) to avoid a large host queue, although you can have higher values when the gateway connects to multiple hosts. That said, the environment must be uniform; do not mismatch the values. For example, if packet sizes exceed MAXDATA or MAXSTL, the packets are segmented into multiple transmissions to the host with significant extra overhead in terms of channel transmission time and processing. When the configuration has multiple communication controllers (that is, the FEPs), set higher congestion priority thresholds. Set MAXOUT=2 or larger to increase the window for multiple transmission without an acknowledgment. This is comparable to the TCP/IP window size or the NetWare PBURST settings.

Outbound traffic is always serialized on gateways, thus, if the DSPU uses 64K RAM adapters, LLC window size should be set to the maximum allowed (maximum receive window size on DSPU, maximum transmit window size on gateway). Inbound traffic might overload a gateway Token-Ring adapter; thus, LLC window size should be set to one (maximum receive window size on the gateway, maximum transmit window size on DSPU). In order to prevent host bottlenecks at the 3174 gateway, you can attach a maximum of two controllers to the same channel. The 3745 communication controller might cause a bottleneck at the TIC or channel connection instead of ALU utilization. It is recommended that you attach only one 3745 gateway per S/370 channel for heavy traffic. When you have virtual routes on 4 Mbits/s Token-Ring, the gateway NIC becomes the first bottleneck as relates to throughput because of packet size differences. You can either equalize the data payload with an overhead hit to the host and FEP, or better yet, install two or more NICs to increase the throughput. For high availability, consider four NICs.

When the Token-Ring traffic is very high because of the host traffic, performance is increased with additional S/370 channels and 3745 channel adapters in parallel. When transfering files (in either direction) between LAN nodes and host, set the LLC frame with a minimum field size of 2K and the LLC window of at least 2. Fine-tune for the most appropriate values. When you have multiple subnets or are routing host traffic between Token-Ring and Ethernet, enlarge the virtual route TG congestion thresholds. With huge traffic load conditions you might need up to 250 or 300K for one priority threshold, and a higher value for the total threshold, for each transmission group used between two communication controllers.

Although I previously discussed the need to configure uniform data payloads and the potential disparity between host, LAN, and remote connections, there are some other important parameters when the LAN clients are connected over a remote WAN gateway. In general, make certain you have enough logical lines available for these connections, or route the transmissions through a LAN-type remote gateway to make many of the host-side problems effectively invisible.

Replication services

When the transaction volume on the enterprise exceeds the available bandwidth, the data can be replicated (copied) and remotely distributed to local servers. This reduces WAN and enterprise traffic, decentralizes database server processing power, and creates redundancy and remote archival possibilities. The key performance improvement is the reduction in the latency for a remote connection. Single-keystroke latency between a remote terminal and host could be as high as five seconds. A user can regain minutes with a local connection to a replicated database. It is also a means of load balancing overloaded database servers. The issue is not so much duplicating data from a primary site to secondary sites, but rather doing so in a selective and sophisticated manner. Oracle, Ingres, Informix, and Sybase are the prime vendors for selective replication services.

The complete database is not fully replicated at all sites; each site has those records most likely to be updated from that site. Changes are then propagated as needed to other servers. This is called *synchronization* and relieves the bandwidth and server messaging burden typical with two-phase commits. The downside to this technology is that it increases complexity, limits integration to a single vendor or platform, and adds significant administration overhead.

Replication services are also important (and a bottleneck) for enterprise network operation. Access to services or devices on the enterprise is often provided through multiple routes. User password and authentication should limit access unless a global directory service (such as NIS, \ETC\HOSTS, ENS, or StreetTalk) is active. When the NIS functions are centralized, the overhead for authentication is significant. However, when the centralized facility is unavailable, access to services or devices is severely curtailed. StreetTalk III (enterprise directory services) provides shadowing and replication of the database. This facility is important (especially for the large networks) because routers frequently access the directory information. Although the overhead for synchronizing the databases might seem excessive—and this is a problem for all routing tables—the background load is actually less than centralized authentication.

Specifically, here are some tips for Lotus Notes network management. Notes services can be optimized by minimizing data propagation and synchronization. Since propagation and synchronization is usually scheduled, the most effective means for minimizing remote site bandwidth, and both local and remote server overhead is to review your replication schedule. Many organizations enable immediate replication of changes, and this is often unnecessary when the receiving organization is a half-a-world away and most workers are sleeping at home. That service is pointless and useless. Replication scheduling is optimized when you understand the *duration* of the replication process. You do not want to overlap replications while an earlier task is still completing. Create a replication schedule that understands the critical requirement for information and the working hours of the receiving organizations. Create concurrent servers or parallel servers to distribute information with either multiple centralized servers or distributed regional servers. Lastly, tailor the data propagation not only to minimize network propagation loading, but also the phone or connection charges.

Replication of data sets is a growing trend for group scheduling, calendaring, and document management activities, as any Notes user understands. The workflow management is critical for maintaining synchronization of these distributed databases. While replication makes information rapidly available locally, many applications are not bullet-proof when it comes to supporting the enterprise environment. Distribute data sets with preset attributes, authentication disabled, or write-access limitations, but create a central clearinghouse to provide local access network-wide. This creates slower enterprise network access for actually scheduling or creating new events, schedules, or documents.

Enterprise naming

Naming devices, users, servers, hosts, files, disks, partitions, services, hierarchies, and other network objects is a challenge in a distributed environment. These names play a key role in security—that is, maintaining security—and security cannot be reliable until you get the naming simplified and globalized. Duplicate names, aliases, and common names for common services are a nightmare on the enterprise network and a performance drain as processes attempt to resolve the logical names with physical addresses. These problems are more acute when the databases are distributed, replicated, and synchronized.

The best way to optimize enterprise naming is to establish a four-tier convention that parallels the TCP/IP naming convention of 000.000.000.000. The highest order represents global naming for resources that transcend the enterprise network and reflect services and information in the real world. This may include external E-mail address, Internet services and places, vendor BBSs, local carrier exchanges, routes, and even physical devices addressed or controlled from the network. If you are parochial and assume your enterprise network is the whole world for your organization, you are limiting growth and locking out access to external databases and services.

The second hierarchy is your enterprise network. Groups, divisions, and other domains should be named and addressed at this second level of significance. The third level is best used for local resources, such as LANs, hosts, and closely managed environments. The fourth and least significant tier is the individual components, such as system names, ports, devices, files, file services, and printers. These names may be fixed by manufacturers or selected by the users. In any event, they should be consistent with the four-tiered global naming convention. You will see significant benefits in terms of ease of management and maintenance, but also significantly faster consolidation of routing addresses, and ultimately, faster throughput and better bandwidth utilization.

Imaging optimization

Document scanning and imaging as well as medical imaging represent significant enterprise network overhead. The requirement for these two very different productions must be recognized. First, the FDA to date does not allow compression of medical images even with the application of lossless or RLE methods; compression is not a legal option. Second, medical images are likely to require 15 minutes to create the 50 to 100 MB image. Third, typical legal or office documents represent about 70 KB, less than one percent of the size of a typical PET, CAT, or digitized X-ray, and these can be compressed and dithered to reduce storage and transmission bandwidth requirements. The techniques for optimizing imaging performance must differ to correspond to the environment.

All imaging operations benefit from staging or pre-fetch of documents. This means that users manually request that images are downloaded to their local workstations if they know that they will need them. Frequently-accessed documents can be cached or stored to faster HSM devices. Although the time required to transmit 1 MB of paged images is substantial on any channel, the transmission time is dwarfed when these pages must be located and transferred from jukeboxes with access time exceeding eight seconds. Although this access time does not seem excessive by itself, the aggregate retrieval time from a WORM jukebox can become hours in a typical law firm or customer claims in an insurance company. Management of the request queue in a optical jukebox is also important. Although first-come, first-served seems the most fair, the time required to switch platters usually makes it more efficient to fulfill currently queued requests from the current platter before fetching the next platter. Parallel jukeboxes with duplicate information bases increase access rates. Measure the $50 to $200 *thousand* jukebox cost against the enhanced performance.

Legal and office document imaging benefit from strategies. Scanned images should be compressed when they enter the system, before they are transmitted over the network and before they are stored. Store them in compressed format and at the highest resolution necessary for display or reproduction. Higher resolution is unnecessary and wasteful of resources. Clean up dirty images when they enter the system. Muddy photocopies, gray-scale renderings of black and white documents, and halftones do not compress as well as clean source images. Additionally, as previously suggested, documents often can be reformatted through OCR, which requires less bandwidth and storage overhead, to reduce overall performance. Documents stored in a text format also increase utility since you can now search and index on keywords or contents rather than just upon the document information created when an image is merely scanned and stored.

Lastly, do not overlook the network structure. Imaging works best if it is isolated from other network activities not only with intermediate nodes, but also when segregated with its own network channels. If bandwidth capacity is a limitation, consider fast Ethernet, FDDI, or a high-speed backplane infrastructure just for the imaging activities.

Printer configurations

Printing over an enterprise network is a substantial resource drain. Recall from Chapter 3 that the typical increases to traffic include large files, graphics, and imaging. Although users on client/server networks typically create many reports and process paperwork that bogs printers, this resource load is relatively small in comparison to graphical presentation and imaging activities. While a large report may consume 75K and require 10 minutes to print all 140 pages on a high-speed chain printer, this nonetheless represents a sparse operation. A single compressed scanned page easily represents 75K in a single page. Furthermore, there are probably another 12 pages to print for every one you think needs printing. The issues in optimizing printer performance are load distribution and load balancing, matching the actual printer unit with performance requirements, and printer management.

Printer load distribution and load balancing is a design and software issue. The most important technique for both improving enterprise network performance and printer performance is to distribute the print load so that most print jobs are not queued and routed over the network. This would represent a gratuitous load. Figure 5.43 shows the effects of creating a report on a processing server, routing the report to a printer server over the network, and then passing that information for a third time over the network for actual printing.

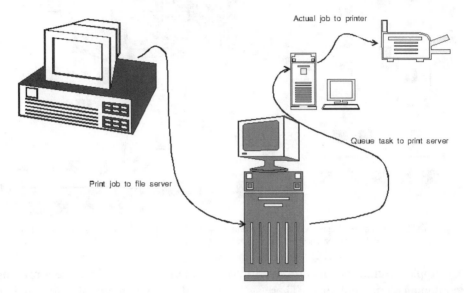

Figure 5.43 A print job is routed over the network channels three times, unnecessarily using available bandwidth.

By redesigning the network so that the processing and printer server are one and the same, the network channel traffic load is totally reduced. Furthermore, the server is spared bus, NIC, and CPU overhead for packetizing and transmitting the report. The only network traffic would consist of messages to the user that a report has been processed, queued, and ultimately printed. However, this may cost CPU cycles, disk space, task swapping overhead, and increase a bottleneck at the server. Bottlenecks at servers can be addressed with faster buses, I/O subsystems, bus mastering controllers, parallel CPUs, and coprocessors. Refer to *Computer Performance Optimization* for other issues about LAN and server architectures. When it is not possible to process and print from the same server, at least avoid routing print jobs over the enterprise. Print jobs tend to be localized and thus should be printed locally. You do not want the report crossing bridges, routers, or gateways, as Figure 5.44 shows, because that only adds a gratuitous load to affected segments.

Output is one of the few results that must be for the convenience of the users; the first available printer or the fastest printer may be physically inconvenient or insecure (for privileged reports). One definition for enterprise network performance optimization may include minimizing the labor for distributing reports and increasing the security of the information. If reports are processed centrally, it may more efficient for the organization to distribute them electronically via E-mail or time-sensitive queues so that reports can be shipped at off-peak times.

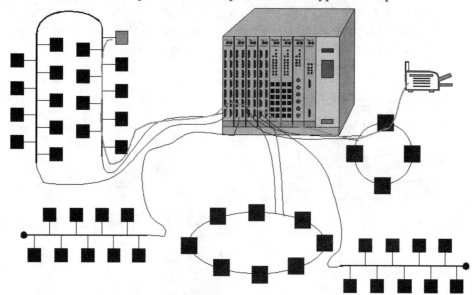

Figure 5.44 Do not route print tasks over the greater enterprise network as is shown here. Performance for all tasks will be better when printing service is local to each segment or subnet.

Apple Macintoshes incorporated into the enterprise network create a printing communications collapse. There is a special access protocol for sending, controlling, and monitoring status of Macintosh printing services that is verbose and bidirectional, unlike most network print jobs. A considerable overhead for the printing job is processed by the client workstation or a special printing server. It is best if this control traffic is routed separately over serial lines. Although this limits communication speed to 38.4 Kbits/s, the channeling represents more efficient use of the enterprise network. Although integration is desirable, the downside of integrating Macintosh print services into the enterprise network is that font information and PDL setup is sent to support jobs in queue. It is efficient for the client and print job, but verbose especially over the enterprise.

Some network operating systems support load balancing on the print queues. Banyan Vines is more advanced than Novell NetWare ENS in terms of balancing print jobs between multiple printers based on current printer status. If one printer is busy, the queue will route the job to another attached that isn't. Priority jobs can be manually pushed ahead of other jobs, or, more intelligently, non-priority jobs can be held.

The last design issue is that of the printer definition language (PDL). A simple ASCII dump requiring 1000 bytes can easily be rendered as a bitmap bloated to 55K or to 75K or more when encapsulated within a PostScript format. Although this may be immaterial to the MS Windows user that selects whatever printer driver seems to work, as you can see, the correct printer driver represents a 90 percent workload reduction.

Using the printer description language efficiently also is important for performance. Although some inefficiency is built into the application selecting the choice of PDL and driving the output—and this gives you little ability to optimize it—how you use the application can optimize the overall effect. Specifically, PostScript Level 1 has *limitchecks,* which limit a path to a maximum of 1500 points. Even though the newer PostScript interpreters handle an unlimited number of points dynamically, compromise and simplify your overly complex curves.

Also substitute color blends rather than calculated gradients. If you must use a gradient, specify the maximum number of bands (that is, resolution steps) that the ultimate output device can actually image. Blends require no CPU or printer memory, and render faster. Combine items that are similar because PostScript will image them at the same time. Do not group them because this requires more memory and calculations to handle the compound object. When items share the same fill, combining them actually requires more processing to match the bitmap to the complex combination.

It is important to match the output device with the requirements of the environment. Install a chain printer or high-speed page printer where necessary. If a printer has a significant backlog, that backlog exists as spool files on disk. Eventually the disk will be near-full with the attendant thrashing problems. When full, that disk will halt processing and possibly create secondary bottlenecks. Most image printers can handle 2 Mbits/s for data transfer, but the parallel or serial ports can handle only 4000 to 30,000 bits/s. The limitation is not in the printer, but rather in the path to the printer.

A Hewlett-Packard LaserJet II is rated at 6 ppm, but typically prints only 1 ppm from a spreadsheet or document. Imaging of a graphic image from a drawing program could require thirty minutes or more. Printers with bus adapters, bidirectional communications, or GDI interfaces (such as LaserMaster, Lexmark, and Microsoft Windows Printing System) can increase data transfer to the printer up to the rated speed. Printers or printer servers supporting emulation of the PDL also represent a bottleneck. You will experience better performance by matching printer drivers, output types, the printers, and queues.

Other industry-standard approaches are also in the works. The fastest that printers convert data into the output image is about 250 Kbits/s—slightly more than the Centronics parallel port speed of about 100 Kbits/s. Also realize that the parallel port cable is effective and reliable only to lengths of about 3 meters. IEEE P1284 is a specification for a faster, standardized operation. It is intended to provide a compatible mode with existing parallel printing interfaces. Nibble mode (4-bit) will provide asynchronous bidirectional transmission, while byte mode provides a wider path. ECP (Extended capabilities port) and EPP (Enhanced parallel port) modes provide an interlocked asynchronous byte-wide transmission channel with separate data and command paths, with optional data compression and multiple data lines. ECP overcomes many of the cable length limitations. IEEE P1394, dubbed FireWire by Apple Computer and TI, allows for up to 63 daisy-chained peripherals from a single I/O port at speeds up to 400 Mbits/s. P1394 uses differential signaling. Although this technology is primarily designed for the desktop, it is pertinent to break the bottleneck caused by printer servers built with desktop hardware and software.

The issue of printer management itself is multifaceted. First, there is the issue of managing a printer that supports multiple PDLs. It may receive plotting, HP, and PostScript jobs in sequence and need to switch between modes. If there is even a remote chance that users will misroute jobs to the wrong printers, you want an automated print facility that will correctly configure the printer, purge the job, or reroute an improperly routed job. Print problems are compounded when the printers are dispersed over the network. SNMP tools (such as NPManager for Windows) provide printer availability information, error types, font type, PDL job type, and paper status and type. The DMTF is in final review of a print management standard that should also provide a means to remotely and interactively configure a printer or correct error conditions. The goal is to manage printers so that they do not create traffic jams.

Servers

A NetWare NLM module called Balance from NetWork Specialists can increase server throughput on single segments attached through multiple NICs. This software provides load balancing as its name implies, and also makes it possible to address multiple NICs in the same NetWare server on the same segment. When used in conjunction with full-duplex Ethernet, throughput can be increased 400 percent. Note, however, that bus mastering cards should be used in this configuration so that the CPU is not fully committed to processing only traffic. The increased throughput also reduces latency as well.

When servers appear overwhelmed at the I/O bus (disk) or at the CPU, SMP-based servers provide a massive I/O channel with CPU and processing scalability. This is the method to bypass file server and database server bottlenecks. Additionally, the architecture and secondary processors usually do not require additional or new software licenses—it counts as one machine. This can be very cost effective. Although most LAN channels will be unable to support the sustained I/O from such a machine, you can partition the server with multiple NICs and blast data out to multiple channels. Also, multiple disks *and* multiple disk controllers improve channel I/O in the server.

NetWare also supports other performance tuning parameters. These include the display relinquish control alerts, maximum allocation for short-term memory, maximum service process time, new service process wait time, and preemptive service time. Refer to the Novell documentation for details; I present this information here because it shows you that there are ways to fine-tune priorities for client/server and application tasks. This is comparable to the PID and process priorities on UNIX tasks.

NetWare 4.x NLMs also provide another degree of freedom with ring settings (that is, application priority). You can increase the priority and hence overall performance of some NLM services to the detriment of others by setting the priority lower (on a scale of 0 to 3). Avoid too many NLMs or fragile ones on the 0-level, as any failures or compatibility problems will crash the server.

Servers typically yield performance with I/O, cache, and disk tuning. Configuration is also an important issue. Tuning becomes more important for mission-critical databases. Create a parallel, scalable structure. Use more controllers and more disks. This does not mean you should not create a single volume (or partition) from these, but performance is typically better this way. Use the raw disk partitions rather than partitions created by the operating system. Place data tables, indexes, views, temporary files, and transaction logs on different physical drives to improve performance. This improves throughput and access time by making more heads available to fetch the data. It also solves the disk-head movement concurrency problem. Stripe data across multiple disks to increase the parallelism. When the database engine supports the *record update in place* feature, you save twice because the record is written directly to disk through the cache to the permanent storage location. The address to the information may still be in cache and the files will not need to be defragmented later. Avoid SCSI-2 controllers. Although the actual transfer bandwidth is greater than SCSI, the drivers and hardware are not yet optimized and provide less performance than SCSI. For additional ideas for tuning servers, see *Computer Performance Optimization.*

As Chapter 3 stated, NFS services are either skewed to file attribute loading or large file transfers. Successful optimization techniques depend on the type of load your enterprise network experiences. If your network is constrained by file attribute searches and small files, the bottleneck is the speed of hard disk access and seeking. This is best optimized with more and smaller disk drives to distribute the seek load over more units. On the other hand, data-intensive sites will benefit most from multiple NICs and faster channels (switched Ethernet or FDDI), or intelligent microsegmentation. NFS version 3 also supports negotiated data-block sizes for block transfers greater than the 8 KB previously supported. The new version supports transfers up 4 GB for triple the sustained data throughput. This fact is valuable not only for NFS but also for other network operating systems. You might want to analyze the disk access distribution; modify sector and block sizes to optimize disk requirements.

Another approach, besides throwing high-end server hardware at the bottleneck, is to replace the NFS server functions with a pared-down operating system, the so-called "NFS appliance" approach. It is typified by FAServer from Network Appliance Corporation. This server only produces NFS services and fulfills directory and file requests; it does not provide general-purpose processing, messaging, or other UNIX services. FAServer improves performance by creating a single file space that spans a RAID subsystem (much like the Stacker for DOS file compression utility). It also creates multiple file snapshots when necessary to resolve access concurrency problems. Performance is generally about 20 percent better than generalized NFS and 50 percent better specifically for write operations because of the spanning volume and RAID optimization.

Many managers regard network and server downtime as anathema. They maintain uptime and availability charts and review the server logs for error and failure messages. Keeping a server up for 98 days without downtime may be a supernatural feat, but one that you should reconsider. There are advantages to bringing down routers, servers, and other network devices to reset the base of allocated resources. The unloading and reloading or NetWare NLMs, temporary memory and background processes for UNIX, RAM, disk caches, and other memory structures do fragment (just as disks do), and sometimes retain very old data. Memory fragmentation is as debilitating as disk fragmentation, but the effects are just that much more, due to the speed differential. While disk I/O may represent 25 percent of device resources (in parallel), memory often accounts for 99 percent of activity. The 10 percent disk fragmentation performance degradation can compare to a 25 percent server performance loss.

Some servers implement error correction code (ECC) to correct memory errors in RAM. This facility is very valuable when static spikes, pulses, electrical noise, or overheating alpha particles (given the small sizes of transistors in silicon) strike critical enterprise network devices. Timing problems between bus and bus mastering cards are common with PC-based hardware. Just because a vendor shows benchmarks that indicate that their adapter will provide greater throughput with less CPU-loading, be certain you confirm these findings in your environment with your hardware and applications.

Radio interference is a problem of degree, proximity, and wavelength. As computer equipment runs faster, does more in hardware, and is built by integrators or even in-house by your organization, RFI and EMI are likely to create difficult-to-diagnose performance problems. They will be acute for mission-critical devices and connectivity devices, and could frequently precipitate the dreaded network panic. Because the FCC Class A and Class B standards only measure radio noise outside the case, not noise inside, you may want to review hardware configuration on suspected components. Also, RFI and EMI may not create outright failures, but soft failures and bottlenecks. This is particularly a problem with faster computers and SMP with multiple CPUs.

Clients

The solutions for improving the performance of a user workstation are detailed in *Computer Performance Optimization* for various operating systems and environments. Rarely will the desktop user be limited by network bandwidth. Because most applications are moving to a graphical interface, the limitation is mostly I/O-based and performance generally can be improved by addressing those bottlenecks specifically.[11] Of course, if I/O or graphics redisplay is not the active bottleneck, adding hardware and new drivers will not affect performance. In summary, these options include:

- Installing a faster local disk
- Cache local and network disks *locally*
- Defragmenting disk files
- Install faster device drivers
- Run nonessential tasks in background
- Add a graphics accelerator or coprocessor

The solutions for terminal communications (SNA and X Windows) bottlenecks are different, as are text-based PCs. The bulk of applications are text-based, and little can be accomplished to improve performance except at the host or server. You certainly do not want to port a slow text-based application into a GUI environment, or translate and encapsulate the original transmission protocol into a different one, as these only slow performance more.

Client communication can be delayed because server resources can be overwhelmed or exhausted by multiple streams and requests from all the other clients. The solution is to adjust the maximum and minimum receive buffers on the *server*. Also, you will want to match the physical receive buffer size to be the maximum packet size plus overhead for decoding these packets. If you are running in a NetWare environment, see "Cache configuration" on page 238.

Many specialty caches, such as printer or CD-ROM caches provide marginal performance boosts. CD-Blitz increases the local access to repeatedly read material on the CD-ROM; this is primarily the volume directory, as you do not really reread the CD-ROM repeatedly for most applications. Read-ahead caching is more effective. Material in cache was returned in 85 μs, in comparison to several-second delays. However, overall utility for a client with a CD-ROM in the enterprise environment is better served by a single cache driver for all local drives, whether tape, floppy, fixed disks, or CD-ROM. Speedcache+ does all that with equivalent performance. This can be very beneficial when you need to off-load overhead from the server, establish a remote office, or want to reduce the WAN traffic. Figure 5.45 shows the control panel for Speedcache.

Typically, though, the bottleneck for the network client is access to the server. CacheAll not only caches local drives, but also network drives and network CD-ROMs. This does reduce the load on the transmission channel and the server by about 5 to 15 percent for client/server applications. Read-ahead buffering of a server is not beneficial for most applications. This merely adds workload to the server, fetching data and shipping it over the network to a client that is not always needed. When read-ahead buffering of server I/O has a hit rate over 50

---

[11]*The Top Five Solutions for Superfast Windows,* Bob Weibel, *PC Computing*, May 1993.

percent, it will save from 300 to 800 ms depending on server and network latency. Reassess it when the server is heavily loaded and hit rates are lower.

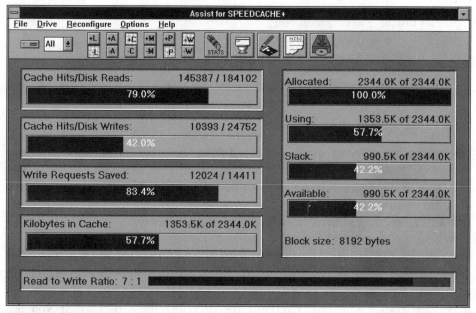

Figure 5.45 The SpeedCache+ control panel.

If clients are running on top of a GUI, understand the difference between preemptive multiprocessing (OS/2, UNIX, and Windows NT) and cooperative multitasking (Windows 3.x). Performance bottlenecks caused by cooperative multitasking can be resolved by shifting to true multiprocessing. However, before you convert your 16-bit applications from Windows into the NT environment, realize that compilation and disk storage will increase by a factor of two, and that 16-bit commands are expanded into 32-bit conversions. Verify that your applications will run faster in the newer environments.

Management of clients on the network is becoming increasingly complex, and it represents a fundamental performance issue because of the financial costs, distance to many clients, and the effects a disconnected client can cause with compound documents. Consider integrating the desktop management interface, plug-and-play, or DTMF agency on your clients.

When installing or upgrading software for many clients, copy the CD-ROM, disks, or tapes to a network file server. Installation will proceed much faster. It is also possible to create one master client installation and copy this disk image to all other clients. Do not overlook configuration files and changes to boot scripts, which typically reside in master directories. Consider modifying the SETUP.INF file that comes with Windows and some Windows-based applications to standardize the desktop interface for all attached network users.

NetWare DOS Requester

Because NetWare resides on some 70 percent of enterprise networks and DOS and Windows comprise the predominate desktop, it is worthwhile considering how to tune the DOS-to-NetWare client interface. This software is called the DOS requester, and some version is required for both DOS, Windows, and Windows NT. Primarily, the performance of the DOS Requester is improved by better management of workstation memory through loading an extended memory manager (XMS or EMS) and attempting to load as much of the DOS operating system as possible in high memory (HMA) with a DOS or third-party memory manager. Although workstation performance is degraded slightly by loading the NIPX, NCP, XNET, or VLM.EXE applications in high memory rather than conventional memory, the overall effect is a net performance gain for the station.

You can improve memory availability by not loading NetWare Directory Services, Bindery Services, or printing services; however, this is usually impractical on the enterprise network, as it limits the user's access to network resources and compounds the effort to manage the network services databases. Most of the performance parameters are contained in the NET.CFG file. This is true for earlier AUTOEXEC.BAT commands too. The workstation has similar parameters to the server cache settings. Unfortunately, there are no utilities yet remotely like Hawknet for setting and tuning workstation parameters. The parameters include:

- Load low CONN
- Load low IPXNCP
- Checksum
- Large internet packets
- Signature level
- Cache buffers (number)
- Cache buffer size
- Commit

To optimize performance for the enterprise network, load CONN and IPXNCP in conventional memory. The NCP checksum is best turned off as this is redundant; the network protocol includes another checksum. Large internet packets increase packet size to greater than 512 bytes, and should be enabled wherever you have gateways or routers. The number of cache buffers needed should match the need for the primary applications. Most DOS and Windows applications require only 5 or 10 Netware buffers, but many complex client/server database applications maintain 40 or 50 open files at a time. Match this to your needs. Similarly, match the buffer size to the protocol in use. Token-Ring should be 4202, FDDI 4586, and Ethernet 1518. ARCNET is at least 611 bytes. The buffer size varies depending upon the level settings for ARCNET. Turn CACHE WRITES=OFF and TRUE COMMIT=OFF so as not to bypass the server caching facility; this will provide substantial performance improvements by actually using the server disk cache instead of bypassing it. Printer caching settings are irrelevant in most environments because printer output is usually a file sent through the LPTx port redirector. It is also convenient to SHOW DOTS=ON for expedient network subdirectory access from File Manager.

Windows Print Manager creates a double queueing as Windows itself queues it before it is redirected to the network print queue. For best performance, enable Print Manager but disable "Fast Printing Direct To Port" from Control Panel. Within the NetWare Print Manager, select the "Print Net Jobs Direct" to bypass any local printing services. Although Windows runs faster *without* Print Manager, overall the user will perceive better system response to print services.

X Windows

X Windows is hungry for bandwidth because of its basic client/server architecture, and the network load can be very substantial when the terminals are diskless. X Windows splits the processing between the server and the client but at a cost in terms of the data that must be moved over the network. On the other hand, a good prescription for client/server bottlenecks is to move more of the client processing to the server with X Windows, as this lowers the bandwidth requirements. X Windows functions within fractional T-1 lines as it is basically a graphic terminal emulator running on TCP/IP. Because all processing under X Windows occurs at the X Server, only screen updates are shipped over the network. If this sounds a little like SNA or a remote control PC, you are right, although the X applications tend to be more graphics- and data-oriented than order entry or financial operations. In other words, you need to compare X Windows bandwidth requirements to comparable client/server applications.

Typical client/server applications run better in an X Windows environment than in a strict client/server processing environment for several reasons. First, processing can be optimized within a single environment, not optimized for server, client, and wire speed effects. Second, it does not require quite as much bandwidth, generally the downfall for client/server applications running over several hops or a low-speed WAN link. Third, performance and reliability is usually exceptional, because the VMS and UNIX operating systems and servers have benefited from many years of performance optimization.

X Windows Version 11 Release 6 (X11R6) and later has an option for serial line connectivity called *Low Bandwidth X,* which optimizes transmission between X Client and X Server by removing unused bytes in requests, performing RLE compression on streams, and sending only changes over the link. This optimization is not appropriate for LAN links, because it trades slow wire speed at the expense of the faster CPU speed. Also, for this technology to work, both client and server must support the enhancement.

Anomalies

Adding RAM can actually decrease performance if the cache and other memory structures are not aware of the reconfiguration, or if the supporting structures are not themselves reconfigured to match the larger memory base. Also some applications, regardless of the environment, platform, operating systems, or compiler, will not address any more RAM than they were designed for. Typically, some OS/2, DOS, and Windows applications will not benefit with any more than 16 MB of memory no matter what you do, with the exception of replacing the workstation or server operating system.

Optimizing WAN connections

The bandwidth performance of broadband SVC, PVC, analog, and digital connections can be improved by multiplexing or compressing the data streams with modems, routers, or other software/hardware combinations that apply a repeating pattern table lookup or run length encoding. Typical compression ranges from 2 to 4 times. In effect, you can double the bandwidth of these transmission channels. The encoding and decoding adds about 100 ms at each end to transmission latency. However, do not attempt to compress data for transmission if it has already been compressed; this creates a significant bottleneck that will increase latency 5 to 10 times.

A multiplexer, or mux, can aggregrate multiple PVC, fractional T-1, T-1, E-1, T-3, or E-3 services to create a single large pipe for traditional LAN-to-LAN loads. Often, you can incrementally add capacity. The limitation is the ability to integrate these high-speed serial interfaces (HSSI) or V.35 serial data interface into the NOS infrastructure. This is easily accomplished for VMS, host, and some UNIX-based environments, but is difficult for NetWare. Vines and NTAS do support serial basic connectivity, and these services are extensible with third-party products. However, as the designers have emphasized basic connectivity over tuned performance, these serial services are often inefficiently implemented. Understand that ongoing emphasis is on NOS and software integration and compatibility rather than performance. As I stated in the beginning of this chapter, scalability is an important design and service criteria, one served very well by a mux-based design. The multiplexer additionally provides on-demand services by interspersing voice with data traffic. Some vendors provide traffic prioritization facilities too, as "Optimizing hub and spoke networks" on page 274 describes. This is often directed to IBM and SNA environments.

WAN connections may be less expensive in terms of the wide-area connection with on-demand SLIP or PPP connections with a bridge rather than a router, which expects the remote connection to be available at all times. Bridge and on-demand dial-up connection provides flexibility of connection which a router link does not. When you are supporting remote connections, consider the performance effects of an actual remote node connection versus a remote control attachment. For more information, see *Computer Performance Optimization*.

If you are specifically running NetWare core protocols, you can optimize the link with products from Novell, Newport System, and Eicon. Not only will some of these products compress the data, they act as routers and filter redundant or extraneous RIP and SAP overheads. It is a particularly good solution for RIP/SAP traffic. The LAN²LAN product also provides good information for matching the CSU/DSU and line speeds with requirements. See also "Optimize transmission costs" on page 203.

Financial performance considerations for WAN links

Although using a modem is easy and often very reliable, it does not provide high-speed connections or good cost-effectiveness for anything more than an hour or two per day. When bandwidth is important, shift to digital services. When latency and call setup times are important, shift to ISDN. See "ISDN call setup" on page 214. Whatever protocol and tarriffed service you select, perform a financial analysis of traffic bandwidth utilization, monthly usage, and costs. See also "Consolidation of transmission services" on page 202. Figure 5.46 below shows that modem and ISDN servers are inexpensive for low utilization, but become relatively more expensive than a PVC as hours or bandwidth utilization increase.

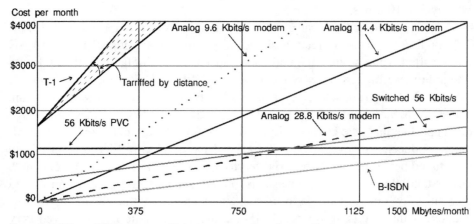

Figure 5.46 Different WAN services provide better economy depending on utilization levels.

Optimizing hub and spoke networks

Boundary routing is the preferred method for attaching remote sites because it is simple. It eliminates the need for the remote sites to download, update, and maintain routing tables. In fact, it needs no routing tables because it relies on the central site for routing appropriate packets over the WAN. Source route bridging is useful for hub and spoke or remote connectivity, but its downside is the overhead for maintaining path and performance control. The distinction that makes a boundary routing device a router rather than a bridge is that it talks to a router at the central site. The limitation of this device is that it only provides for single point-to-point connection. Boundary routing sometimes provides a built-in redundancy, so that when a dedicated leased line fails, a dial-up connection can be established to maintain connectivity. Some vendors also use this facility to support high bursts of LAN traffic. When the dedicated line is overextended it can be supplemented with the extra dial-up bandwidth. Your situation may require more complex connectivity to support SNA traffic as well as LAN traffic. You might consider the efficiency of multiplexing voice, fax, and the data streams into a single pipe (as with Hypercom) to gain synergy from a larger WAN link.

ISDN is a good on-demand WAN link. It is less expensive than a leased line with comparable bandwidth, as the prior graph illustrated. The basic ISDN service is BRI, but it is expandable to PRI. Furthermore, because it is a switched connection, you do not pay connection charges when it is not active. The ISDN call setup is 250 ms for local connections, but always less than 3 s. This makes it useful for establishing temporary connections, such as connections to handle bursts of LAN traffic over a WAN. An inverse multiplexer (such as the Telebit NetBlazer) can allocate bandwidth-on-demand within the limits of the channel.

Instant MHS services
If your organization requires the enterprise network for mail and file transport, you do not have to build your own network infrastructure for these services. Global MHS (Mail handling services) can send E-mail and attached documents over CompuServe for $8.95 per month plus $0.22/min for MHS connection time. This service is world-wide and considerably more flexible than expanding the enterprise network with WAN connections, modems, security facilities, and management. Servers can automatically connect to CompuServe and poll for new mail on an hourly basis. This is less expensive than any homemade network, very reliable, and integrates well with Novell, DaVinci, cc:Mail, Microsoft Mail, and other standard messaging software.

Incidentally, you are not limited to just CompuServe. Internet, MCI Mail, Western Union, and more service providers that blast mail and faxes to mailing lists represent other performance optimization options. In fact, some service organizations can compare your addresses against a known list of E-mail addresses and send only those that are fax-accessible via this more expensive route. E-mail is 9600 chars/m vs. high speed facsimile which represents about two pages/m, about 3600 chars, or 150 MB of image. Furthermore, some service organizations can also send thousands of facsimiles to a mailing list in a matter of minutes. This type of communication is often more reliable and more economical than internal extensions to your network.

Improve local and remote performance
The routing protocol used for a hub and spoke configuration is usually immaterial so long as traffic is contained on the originating segment or subnet; only

traffic specifically directed to other segments or subnets should be routed. This is a network-level (ISO level 3) or transport-level (ISO level 4) procedure. However, one routing protocol, the boundary routing protocol, is particularly efficient as mentioned in "Optimizing hub and spoke networks" on page 274. This is also called perimeter routing. However, as stated in "Routing algorithm" on page 218, avoid the source routing bridging algorithm, as this will create an excessive amount of traffic sent out across the WAN links.

Modem speeds are improved by working with the LECs and long-distance services. Consider creating an environment with modems from the same manufacturer. Just because a modem is Hayes-compatible does not guarantee that it will talk to other (but different) such units. When modem (analog) speed and quality become a limiting factor, consider ISDN or other ditigal services. Not only are they faster, the call setup is faster.

As Chapter 3 explained, modem transmission speeds also bog down with old UARTs, software accessing the DOS INT14 interrupt, and software that itself is not as fast as the serial ports and modems can sustain. One performance tip is to try different remote access software. The NASI (the 6B Interrupt) and NCSI, which redirect serial communications over the LAN to a port on a communications server, do not need to check with the DOS vector before transmitting every character. This is an issue for Vines, NetWare, LANtastic, LAN Server, and LAN Manager NOSs. It is not a problem for NFS or NTAS. Similarly, TurboCom/2 bypasses this INT14 and replaces Windows 3.x serial port drivers to activate the 16550 UART for multitasking throughput of up to 19.2 Kbits/s and 50 Kbits/s with compression.

Here is a new technique, the use of the *parallel* port rather than the serial port, with a V.34 modem (115.2 Kbits/s) for high-speed connections. Just as LapLink and other downloading software have shifted from null-modem serial connections to the faster parallel connections, so too has other remote control software. Parallel ports typically run about 5 times faster, and print to high-speed printers about 40 times faster on graphics- or data-intensive jobs because of a better interrupt mechanism and the 8-line parallel connection, rather than the two-wire serial connection.

When running applications designed or seasoned on 10 or 16 Mbits/s LAN links over remote links, you need to understand how they will behave at modem, fractional T-1, or frame relay speeds. You may best optimize enterprise network performance by getting rid of these applications. Although there are many previously mentioned ways of squeezing more bandwidth from WAN links with different services, remote control, compression, replication, local file services, and bandwidth-on-demand, you may find that the greater good is best served by not pushing full remote access over the WAN.

Conclusion

To improve performance, you have to understand the bottleneck. A computer system and its applications represent a complex system of events, data routing operations, and component performance capacities. Single events or single components can create a gridlock that isn't easily perceived. Furthermore, it is possible to misrepresent the application and how it works, what it does, and the options it provides. Sometimes the bottleneck is how an application is used. The wrong modes, commands, and techniques can spawn an excessive load when other methods will yield the same results faster. And sometimes the results are not even the correct results. Make certain, first, that you understand the process and the different paths you could take to achieve your results. You will achieve better optimization when you understand the interactions and have more options.

When it is clear that the options are limited, the performance of applications is best enhanced by either improving the application itself or switching software. Move from resource-intensive GUI environments to text-based applications. Critical applications and ones that are not easily replaced present a more complex problem. If possible, optimize performance by throwing hardware at the bottleneck. Know the bottleneck. Understand the effects of increased RAM. Add more RAM for larger CPU caches. Increase the size of disk caches. Defragment hard disks and increase the amount of free storage disks. Add hard disks or long-term storage media supporting faster seek and transfer times. Replace simple disk controllers with faster interfaces, or add multiple disks. Replace a slow system with one with a wider or faster bus, a faster CPU, and a faster video display. When it seems that multiple components are creating overlapping bottlenecks, avoid the trap of upgrading a computer system in small pieces and small steps; the time can be better spent evaluating the bottlenecked system and entirely replacing it with a faster platform or application.

Chapter

6

The CD-ROM

Introduction

Anyone who confronts enterprise network design, installation, operations, and management is typically overwhelmed. You will not have time to view every demonstration, review the promotional material with its emphasis on making the product look its best, or sort through the dreck for the middleware you need. If nothing else, this book tries to steer you away from marginal or hyped approaches to performance optimization and show you the most efficient and simplest methods to get better performance, increased security, and reliability, while still keeping the network together. Therefore, the included reference CD-ROM has only information that is pertinent to your needs. It includes this book in searchable hypertext format, and utilities for modeling and mapping the enterprise.

A CD-ROM stores about 660 MB on a read-only medium. CD-ROM was selected in place of 5¼-inch or 3½-inch floppy diskettes because of its much larger storage density. The material on this disk can be read on any MPC-compliant CD-ROM reader, and most of the material is designed to be used in DOS or Windows. This disk can be installed for access on a network CD-ROM jukebox for sharing, but it is licensed only for single reader use at one time.

Handling

The disk is impervious to magnetic fields and most hazards. It should not be bent, subjected to heat, used if warped, or its underside exposed to the acids and oils on your fingers. Handle carefully by the disk edges or top surface. When not in use, store in its case either upright or flat.

CD-ROM contents

The material on the enclosed tools CD-ROM includes an MS Windows help file, a multimedia version of this book, performance benchmarks for various operating systems, and some specialized optimization tools. Most of this material is freeware, demonstrations, shareware, or vendor material. All is copyrighted and may not be copied or distributed unless the vendor specifically allows redistribution. Installation and the usage for each tool on this CD-ROM is explained in this chapter. If there is a license or registration fee for shareware, follow the vendor instructions for registration. The CD-ROM tools directory is shown in Figure 6.1.

```
Volume in drive E has no label
.                <DIR>        04-04-94    7:55a
..               <DIR>        04-04-94    7:55a
LAST_MIN TXT       2037 10-01-94   12:00p
COMNET           <DIR>        07-18-94   11:06p
GRAFBASE         <DIR>        07-18-94   11:06p
INTRAK           <DIR>        07-18-94   11:06p
LANBUILD         <DIR>        08-15-94    5:10p
LANCAD           <DIR>        08-15-94    5:17p
LZFWINST         <DIR>        07-18-94   11:04p
NETFX            <DIR>        07-18-94   11:04p
NETSPECS         <DIR>        07-18-94   11:05p
NETVIZ           <DIR>        07-18-94   11:05p
PROPHESY         <DIR>        07-18-94   11:04p
RUMMAGE          <DIR>        07-18-94   11:06p
SIMSOFT          <DIR>        07-18-94   11:05p
SYNOPTIC         <DIR>        07-18-94   11:04p
SYSDRAW          <DIR>        07-18-94   11:05p
WWW201S          <DIR>        07-18-94   11:04p
APPLAUSE WAV      42732 10-04-94    4:26p
BANG     WAV      13138 10-04-94    4:26p
CHARGE   WAV      40628 10-04-94    4:26p
CLAP     WAV      30764 10-04-94    4:26p
D2HNAV   EXE     241712 08-06-94    5:05p
ENETPERF DHN      15860 09-27-94    9:20p
ENETPERF HLP   13675849 09-29-94    7:16p
ENETPERF MVB   13119042 10-04-94    2:54p
GLASBREK WAV      35624 10-04-94    4:27p
IGNITION WAV      27808 10-04-94    4:27p
ILOVEU   WAV     180148 09-29-94    8:24a
LANBUILD EXE    2355648 09-16-94    5:37p
LANCAD   EXE    3819987 09-16-94    5:38p
LANMODEL EXE    2881179 09-16-94    5:38p
LAST_MIN TXT       1661 09-30-94    9:58a
MUSIC    WAV      11662 09-26-94    6:30a
MVIEWER2 EXE     286704 03-22-93   12:00a
RACEHORN WAV      12626 10-04-94    4:28p
SOURNOTE WAV       3596 10-04-94    4:28p
SYSOPT   DOC        703 01-21-94    6:31p
SYSOPT   HLP      79955 12-21-93    9:06a
TRAFFIC  AVI   79190964 09-29-94    1:43p
TUNING   AVI   12219252 09-29-94    2:07p
WELCOME  WAV      13450 10-04-94    4:29p
WHATUDO  WAV     117918 09-29-94    6:37p
        27 file(s)    20529212 bytes    74848 bytes free
```

Figure 6.1 File contents of the reference CD-ROM.

Installation

Most of the material on this CD-ROM is stored in a compressed format to save space, maintain distribution integrity, maintain vendor assurance for proper installation, and provide the installation process with correct paths, configuration, and automatic group and icon setup. Unless specifically noted in this chapter, most applications install from Windows Program Manager with a SETUP.EXE run command. Some MS Windows executable code can be run directly from the CD-ROM by double-clicking on the filename in File Manager, or by typing the full path and filename from the File/Run menu item in either Program Manager or File Manager. DOS applications can be run within a DOS compatibility box under MS Windows, OS/2, or UNIX. You may also be able to double-click on DOS programs under other GUIs from your Program Manager or File Manager and have them execute if DOS support is enabled through the native GUI (this does include MS Windows, OS/2, and SunOS). OS/2 programs will not run except under OS/2. MS Windows applications typically will execute under OS/2 2.2, OS/2 for Windows, SoftPC, or WABI compatibility boxes.

On-line hypertext

The CD-ROM contains the full text and graphics of this book in two versions. The first is in Windows hypertext help format, while the second is in Microsoft Viewer format. The primary difference is that the Viewer format requires the Viewer run-time engine, but it also provides the means to search on keywords of your choice. You can run the hypertext or viewer format directly from the CD-ROM, although it will perform better from a hard drive. The next two sections detail usage of the multimedia viewer and hypertext help versions.

Windows help format

Click on the help file, ENETPERF.HLP, from within File Manager or run WINHELP.EXE and select this file. Within a few moments you should see the opening screen, as displayed in Figure 6.2:

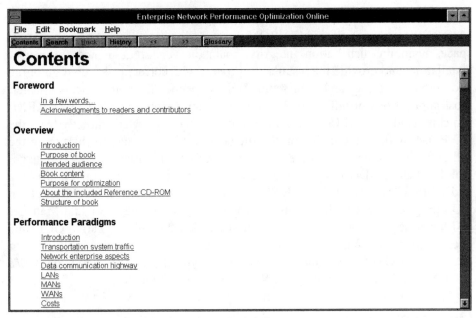

Figure 6.2 File contents of tools CD-ROM Windows on-line help file.

When you click on an underlined entry, in this case <u>LANs,</u> you will access the section on LANs, as shown in Figure 6.3.

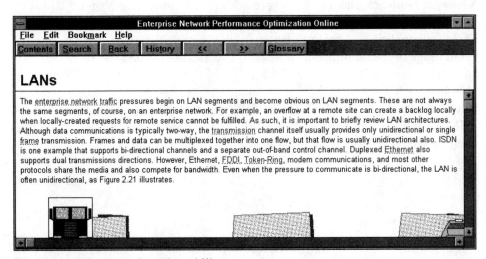

Figure 6.3 A hypertext jump to the section on LANs.

When you click on an underlined entry from within a subject, it refers to a cross-reference or popup definition. Here, selection of <u>FDDI</u> offers the secondary selection of <u>Fiber Distributed Data Interface.</u> When you click on that underlined phrase, you will get the complete popup definition, as shown in Figure 6.4.

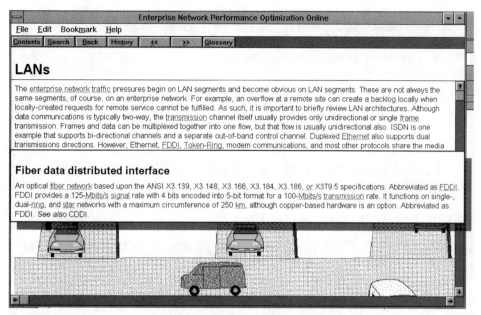

Enterprise Network Performance Optimization Online

File   Edit   Bookmark   Help

| Contents | Search | Back | History | << | >> | Glossary |

LANs

The enterprise network traffic pressures begin on LAN segments and become obvious on LAN segments. These are not always the same segments, of course, on an enterprise network. For example, an overflow at a remote site can create a backlog locally when locally-created requests for remote service cannot be fulfilled. As such, it is important to briefly review LAN architectures. Although data communications is typically two-way, the transmission channel itself usually provides only unidirectional or single frame transmission. Frames and data can be multiplexed together into one flow, but that flow is usually unidirectional also. ISDN is one example that supports bi-directional channels and a separate out-of-band control channel. Duplexed Ethernet also supports dual transmissions directions. However, Ethernet, FDDI, Token-Ring, modem communications, and most other protocols share the media

Fiber data distributed interface

An optical fiber network based upon the ANSI X3.139, X3.148, X3.166, X3.184, X3.186, or X3T9.5 specifications. Abbreviated as FDDI. FDDI provides a 125-Mbits/s signal rate with 4 bits encoded into 5-bit format for a 100-Mbits/s transmission rate. It functions on single-, dual-ring, and star networks with a maximum circumference of 250 km, although copper-based hardware is an option. Abbreviated as FDDI. *See also* CDDI.

Figure 6.4 Hypertext jump to a popup definition, FDDI, and then to Fiber Distributed Data Interface.

If you click on the search command button from the help screen, you will see a popup window as shown in Figure 6.5. Select a topic and double-click to see the options available. As the topic for Balance NLM shows, there are two major references to this topic in the on-line help. You can select either one. Some topics have twenty or more references.

You can create annotations (your own additions to the on-line text file) by selecting the annotate command button from the help menu. When you create an annotation, you will see a paper clip icon in the help file. You end the help session by ALT-F4 or File/Exit key sequences.

Multimedia Viewer format

The on-line Windows help format is very powerful, but it lacks the keyword and full-search text facility and native support for 256-bit color graphics which the more powerful Windows Multimedia Viewer provides. I have included the book in the Viewer multimedia format primarily to take advantage of the keyword search feature. To initiate Viewer, run the viewer file MVIEWER2.EXE, and you should see the opening screen as shown in Figure 6.6.

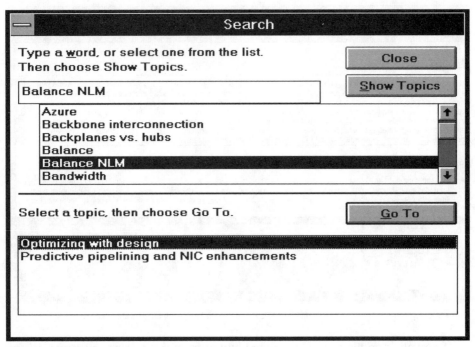

Figure 6.5 Search of Balance NLM yields two references to this topic.

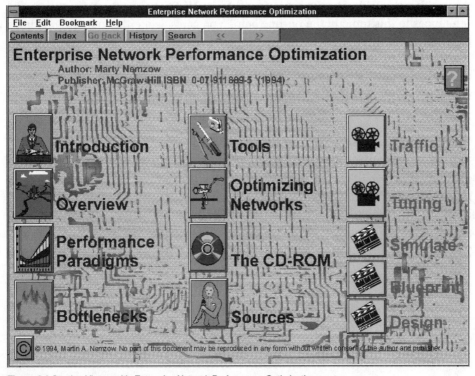

Figure 6.6 Starting Viewer with *Enterprise Network Performance Optimization*.

The process to load the ENETPERF.MVB file for the *Enterprise Network Performance Optimization* title is this: type MVIEWER2 E:\ENETPERF.MVB. Substitute the drive letter corresponding to the CD-ROM on your system. If you have established a file extension association for MVB files in File Manager, you can also double-click on the ENETPERF.MVB filename. When you select the Introduction, you should see a window as illustrated in Figure 6.7.

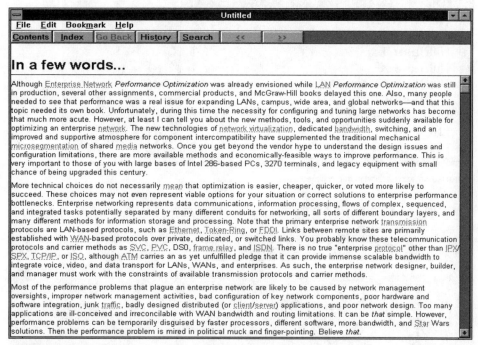

Figure 6.7 Loading the multimedia version of *Enterprise Network Performance Optimization*.

Viewer navigation is very similar to the Windows on-line help. If you are comfortable with standard Windows help navigation, you will have no trouble with Viewer. In fact, you will probably like it better. In addition to the standard hypertext popups, there are graphical hot spots, a few more command buttons and hypertext jump options, all of which are generally self-explanatory. The index popup and scroll list provides immediate jumps to chapters, sections, and key topics. Figure 6.8 shows these options.

The screen shot in Figure 6.9 illustrates not only the keyword search feature in Viewer, but also the search result scroll list of related topics. This is the fundamental functional difference between Windows on-line help and the multimedia format. Enter your keyword and Viewer will locate all instances and provide an occurrence count. This is important because some keywords are too common to reveal a narrow search.

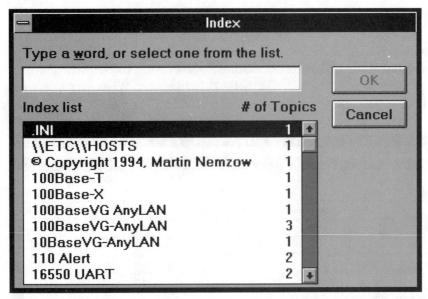

Figure 6.8 Navigating through the multimedia title.

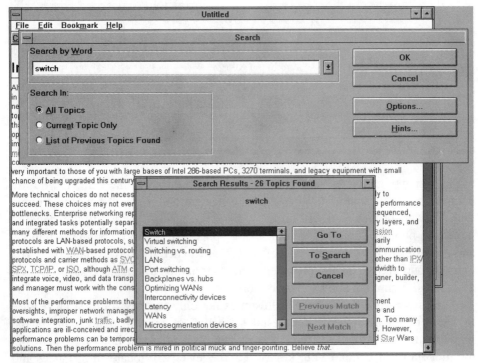

Figure 6.9 Searching for keywords and topics within Viewer.

Computer Performance Optimization demonstration

A limited version of _Computer Performance Optimization_ was authorized for distribution on electronic bulletin boards (BBSs) as SYSOPT.HLP. SYSOPT. DOC is a description of the contents of SYSOPT.EXE. This file, which is in Windows hypertext format, is invoked by double-clicking on its file in the File Manager, or you can drag the file from File Manager into Program Manager to create an icon. When you double-click on the filename or its icon, you should see the sizable screen shown in Figure 6.10.

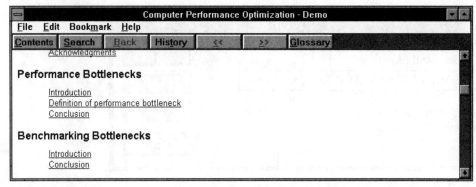

Figure 6.10 Part of the main help screen for the BBS edition of _Computer Performance Optimization._

Demonstrations

The demonstrations provided on the CD-ROM include:

- GrafBase
- SysDraw
- Synoptics 3174 workstation hub module
- NET F/X
- ServerTrak and TrendTrak
- LANalyzer for Windows
- Prophesy
- PC Model
- CACI
- SimSoft
- LANBuild
- netViz
- Win, What, Where
- LAN•CAD

Each section details the installation process and then shows how to begin running the slide show or working model.

GrafBase demonstration

GrafBase is a network diagramming tool for showing the location of sites, connections between sites, and documenting configurations. The demonstration is a DOS-based slide show. The file, SETUP.BAT, expands the executable code with embedded graphical images. You can exit to DOS or execute this file from within a DOS window (compatibility box). Alternatively, you can install the software from Program Manager, as shown in Figure 6.11. You must specify the destination drive for the 1.5 MB of expanded files in any event.

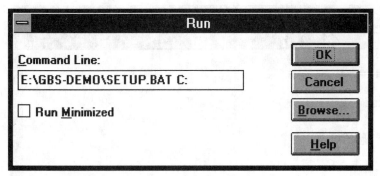

Figure 6.11 Installation of the GrafBase show.

This demonstration can be run under DOS or under a full-screen display in Windows; you really cannot run it well under a compatibility box because it expects full access to the keyboard and the graphics adapter. You invoke it by running DEMO.BAT. You should see screens that look like Figure 6.12.

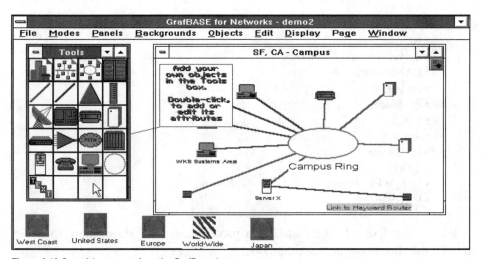

Figure 6.12 One of the screens from the GrafBase demo.

If you receive a warning message about .PIF files and screen device drivers, the inability to access the screen display, or screen resolutions, press [ALT] [ENTER] to create a full-screen Windows display for the DOS compatibility box from the tiled window. If you need to, you can exit the show with the [ESC] key.

NET F/X

NET F/X is a network data collection, testing, and modeling tool for large networks. The demonstration is a DOS-based slide show. The file, INSTALL. BAT, expands the executable code with embedded graphical images. You can exit to DOS or execute this file from within a DOS window (compatibility box). Alternatively, you can install the software from Program Manager, as shown in Figure 6.13. You must specify the destination drive for the 4.5 MB of expanded files in any event.

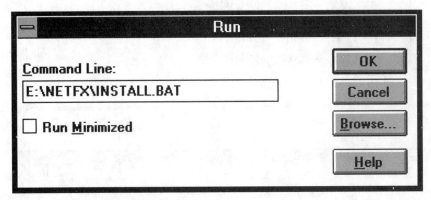

Figure 6.13 Installation of the NET F/X slide show.

You will be asked to specify a destination drive with sufficient capacity for expanded files. This demonstration can be run under DOS or under a full-screen display in Windows. You invoke it by running NET-FX.BAT. You should see screens that look like Figure 6.14. If you see a colorful but scrambled display, or receive a warning message about .PIF files and screen device drivers, the inability to access the screen display, or screen resolutions, press [ALT][ENTER] to create a full-screen Windows display for the DOS compatibility box. By the way, you exit the show with the [ESC] key. Net F/X also has significant charting capabilities, as Figure 6.15 illustrates.

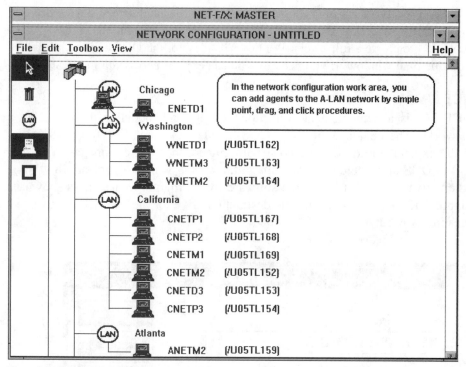

Figure 6.14 One of the screens from the Net F/X demo.

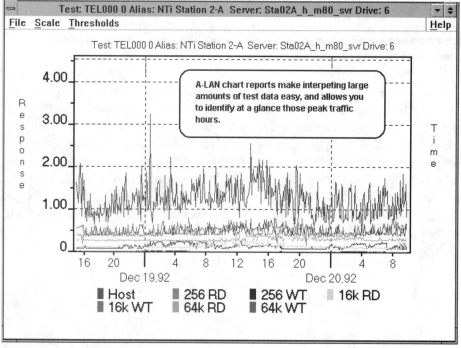

Figure 6.15 One of the charts from the NET F/X slide show.

Synoptics SNA workstation controller

As stated in Chapter 5, one option (depending upon your workload and latency requirements) for optimization performance is to integrate LAN and SNA/SDLC traffic into the enterprise network infrastructure. Synoptics has a nice DOS-based multimedia movie showing how the jointly produced IBM and Synoptics hub card can off-load or replace a 3174-series front end processor.

You can invoke the demonstration directly from the CD-ROM by double-clicking on 3174DEMO.EXE (in the \SYNOPTIC subdirectory of the CD-ROM). This will invoke a full-screen DOS compatibility box for the demonstration. Use the arrow keys or click on the highlighted push button with your mouse to advance the slides. You can end the multimedia show with File/Exit at any time. Eventually, you will see a screen similar to Figure 6.16.

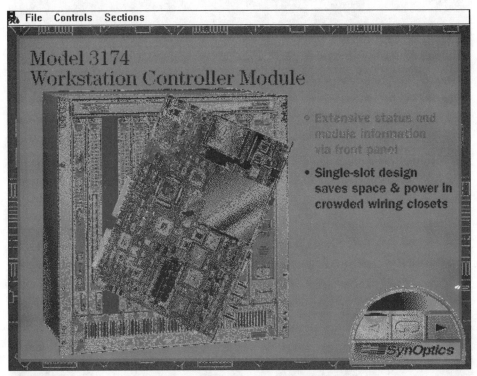

Figure 6.16 A static view of the Synoptics controller card replacing an IBM 3174 FEP.

SimSoft

PC Model is a DOS-based process modeling tool, and the demonstration is a menu system that runs a series of Aldus Persuasion slides and invokes a limited version of the actual modeling software. The easiest way to run the demonstration is to create a Windows full-screen DOS compatibility box and type "DEMO.BAT." Alternately, double-click on the DEMO.BAT filename in the \SIMSOFT directory on the CD-ROM. If you receive a warning message about

.PIF files and screen device drivers, the inability to access the screen display, or screen resolutions, press [ALT][ENTER] to create a full-screen Windows display for the DOS compatibility box from the tiled window. If you need to, you can exit the show with the a combination of the [F10] and [ESC] keys. The first screen is shown in Figure 6.17.

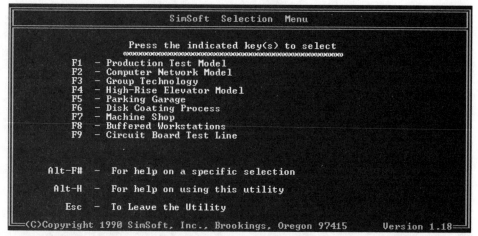

Figure 6.17 The main menu screen for the SimSoft PC Model working demonstrations.

There are 14 working models already established. You can select any one and modify the values, parameters, and time scales. There is one model specifically geared for data communications. If you select [F7] and then [F2], you should see this model for a three-node terminal and host "network" shown in Figure 6.18.

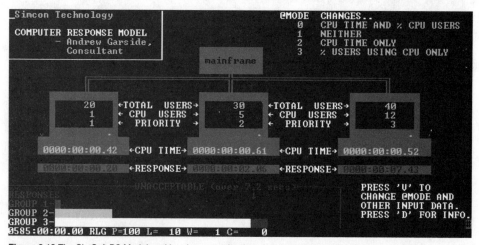

Figure 6.18 The SimSoft PC Model working demonstration for a terminal communications environment.

You can alter the values and add devices to this model, but you cannot save your changes. The value screen in Figure 6.19 shows some of the parameters.

You can also invoke the working model by typing the filename or double-clicking on PCMODELD.EXE, which starts a full-screen DOS session, starts the modeling software, and bypasses the menuing system.

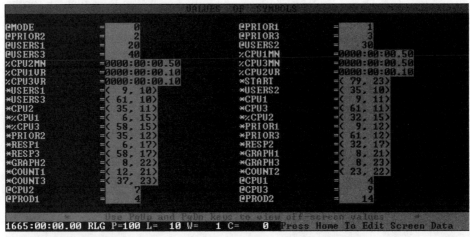

Figure 6.19 The working parameters for the terminal communications environment model.

CACI COMNET III

COMNET III is a redeployment of COMNET II.5 and the L NET II.5 tools. The included demonstration runs a series of video slides that show the capabilities of the actual modeling software. The easiest way to run the demonstration is to create a Windows full-screen DOS compatibility box and type "SETUP.BAT." The first installation screen is shown in Figure 6.20.

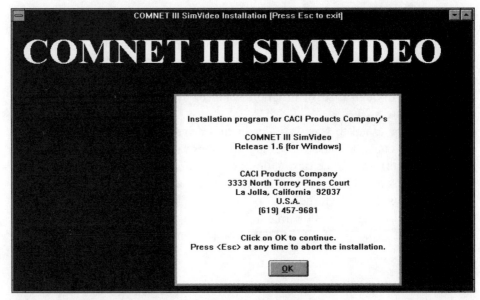

Figure 6.20 The main menu screen for the COMNET demonstration slide show.

Alternately, double-click on the SETUP.BAT filename in the \COMNET directory on the CD-ROM. If you receive a warning message about .PIF files and screen device drivers, the inability to access the screen display, or screen resolutions, press [ALT][ENTER] to create a full-screen Windows display for the DOS compatibility box from the tiled window. If you need to, you can exit the show with the combination of the [F10] and [ESC] keys.

Follow the installation instructions to install it to a directory on a hard drive of your choice (as the video runs too slowly from a CD-ROM to have been included in the expanded format on the CD-ROM.) Upon completion, you will see a Windows icon and group, as shown in Figure 6.21.

Figure 6.21 The COMNET demonstration group and icon.

Click on the COMNET icon to start the video. You might note that this is not a particularly well-behaved Windows application, in that you cannot review or fast-forward through the sequence as with some of the other demonstrations. It is full-screen, does not coexist well with other background or iconized tasks, and creates some windows that take priority over all others. However, you should see a screen much like this, as shown in Figure 6.22.

ServerTrak and TrendTrak

These two demonstrations are working versions of Novell server performance monitoring tools. Additionally, on-line help provides a good overview for using this tool, and some interpretation of the sample data sets. Installation is accomplished through DOS or Program Manager by running INSTALL.EXE, as shown in Figure 6.23.

You will be asked to specify a destination drive. The setup application will expand and copy all the necessary demonstration files, and executable ServerTrak (lite) files to the destination drive, search for Novell server information, and do some housekeeping. Because these are fundamentally DOS and server applications, you will need to invoke them from a DOS prompt. Also, you will probably need at least 500K free in DOS to run the slide show demonstration of ServerTrak. Type DEMO at the DOS prompt to invoke it.

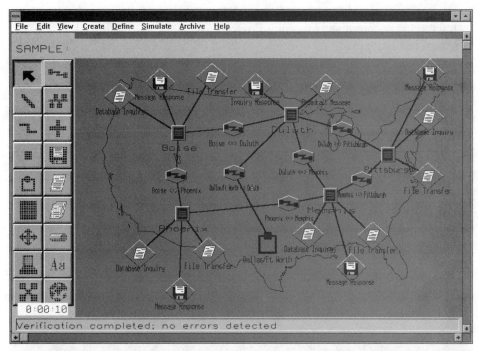

Figure 6.22 A screen shot from the COMNET video showing the message (blue box at Fort Worth) moving between network sites.

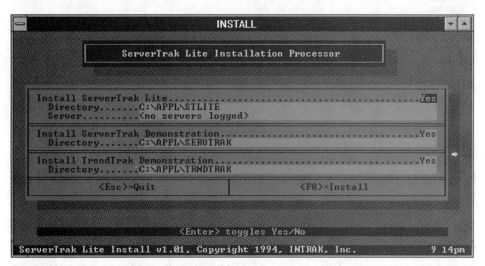

Figure 6.23 Installation of the InTrak software is a DOS process.

LANalyzer for Windows

This demonstration is a working version of the Novell product, without working data capture facilities. There are several data sets included, with notes explaining how to use the protocol analyzer and correctly debug the constructed performance problems in the data sets. Additionally, on-line help provides a good overview for using this tool and some interpretation of the sample data sets. Installation is accomplished through Program Manager by running SETUP.EXE, as shown in Figure 6.24.

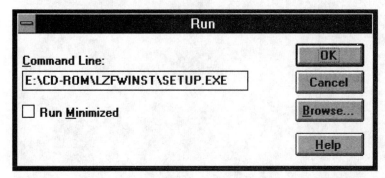

Figure 6.24 Installation of the LANalyzer for Windows demo is through the standard Windows process.

You need to specify a destination drive. The setup application will expand and copy all the necessary files, do some housekeeping, and create a group in Program Manager, as shown in Figure 6.25.

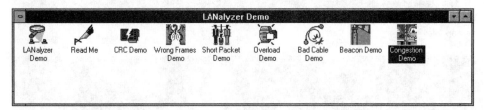

Figure 6.25 LANalyzer for Windows installation creates the icon and group automatically.

The installation process also creates several multimedia Toolbook applications for training and demonstration purposes. The LANalyzer demo (first icon in the group) runs a Toolbook multimedia show. The other icons, as shown in Figure 6.26, invoke a working version of LANalyzer for Windows. The housekeeping for these samples are actually fairly extensive because the demo dummies up the necessary drivers for either Ethernet or Token-Ring. These require TSRs to be loaded *before* invoking Windows. Note that the modifications for AUTOEXEC.BAT, CONFIG.SYS, WIN.INI, and SYSTEM.INI are easily reversed with the uninstall option in the SETUP.EXE application. You may want to read and review the .WRI and README files to check for proper installation. To invoke one of the models, double-click on the icon of your choice. Activation is subject to initialization of the correct protocol configuration TSRs in the DOS boot files and Windows configuration files; the LZENET.BAT must be run for the Ethernet samples and LZTNET.BAT for the Token-Ring samples. LZEND.BAT will unload the TSRs.

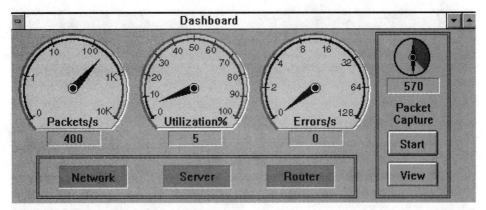

Figure 6.26 A screen shot from LANalyzer for Windows.

Prophesy

Prophesy is queueing model that will model almost any heuristic process. The Prophesy demo is a DOS-based slide show of Windows screen shots with notations and explanations. The easiest way to run the demonstration is to create a Windows full-screen DOS compatibility box and type "DEMO.BAT." Alternately, double-click on the DEMO.BAT filename in the \PROPHESY directory on the CD-ROM. If you receive a warning message about .PIF files and screen device drivers, the inability to access the screen display, or screen resolutions, press [ALT][ENTER] to create a full-screen Windows display for the DOS compatibility box from the tiled window. The left and right arrow keys advance or recall the prior slides. The [ESC] key will exit the show.

SysDraw

Microsystems Engineering's SysDraw is a general-purpose drag-and-drop illustration package (like CorelDraw or Visio) with a wonderful set of clip-art specific to the creation of computer and network diagramming. The demonstration is a Windows-based automated demonstration of the tool's capabilities. Invoke the demonstration by double-clicking on the filename SYSDRAW.EXE (in subdirectory \SYSDRAW on the CD-ROM). The demonstration runs automatically in the background, but it requires a full-screen window. You might note that this demonstration runs faster than the drawing actually does. Also, I imported several objects into CorelDraw through the clipboard to see if I could cut and paste a figure from the SysDraw library. The images seem to be EPS-based and decompose into 600 to 900 objects, which is more graphical resources than many other illustration packages can track. Figure 6.27 illustrates the SysDraw demonstration.

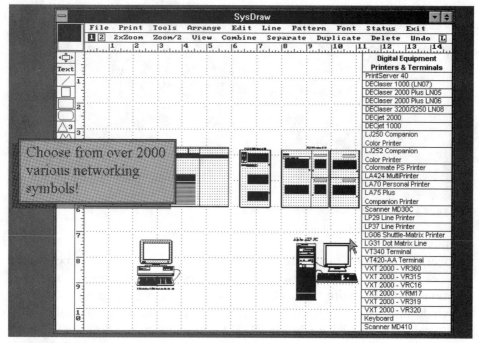

Figure 6.27 The automated SysDraw demo shows off its capabilities.

LAN•CAD

LAN•CAD is profiled in Chapter 4 for its ability to create llustrations of the network for the bill of materials and queueing modeling tools. This is a shareware demonstration and a limited version of the full commercial application. You are allowed to pass the software demo along subject to copyright restrictions. Installation is accomplished through Program Manager by running SETUP.EXE, as shown in Figure 6.28.

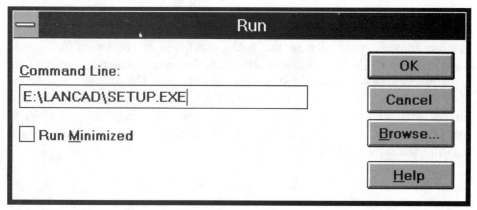

Figure 6.28 Installation of the LAN•CAD demo is through the standard Windows setup process.

You can use the program to print or create screen shots of larger networks. The full installation requires about 0.75 MB. You have the option of installing to a directory of your choice, as Figure 6.29 shows.

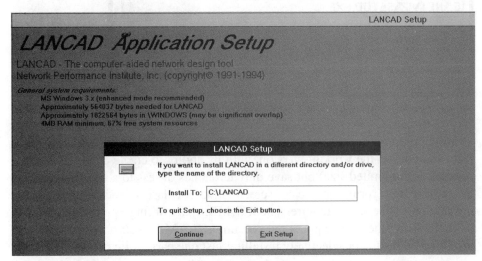

Figure 6.29 You can choose to install the software to a directory of your choice.

Setup creates the Program Manager Group and icons, as Figure 6.30 shows.

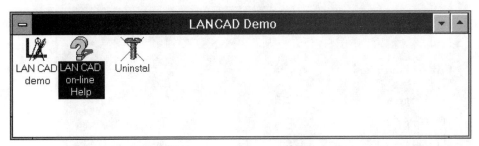

Figure 6.30 LAN•CAD installation creates the icon and group automatically.

To invoke LAN•CAD, click on the icon. The demo limits you to network designs with no more than 20 items or nodes; this includes nodes, servers, wiring, and any blueprint items such as doors, rooms, and lines.

netViz

NetViz is profiled in Chapter 4 for its ability to create "drill-down" illustrations of the network and create flow diagrams. This is shareware demonstration and a limited version of the full commercial application; there are two parts to the demo. You are allowed to pass the software demo along subject to copyright restrictions. Installation is accomplished through Program Manager by running SETUP.EXE, as shown in Figure 6.31.

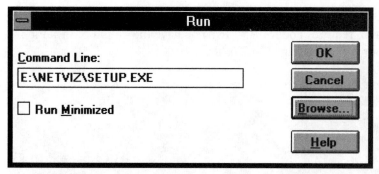

Figure 6.31 Installation of the two-part netViz demo is through the standard Windows setup process.

Installation options include an automated show tour of the application and its features and also the working version, with its ability to create a large enterprise network (of unlimited size) but save only ten nodes or less to a file. You can use the program to print or create screen shots of larger networks. The full installation of the demo requires about 5 MB, the working application requires 2.8 MB, while the tour requires the remaining 2.2 MB. You have the option of installing both parts or selectively installing just one part, as Figure 6.32 shows.

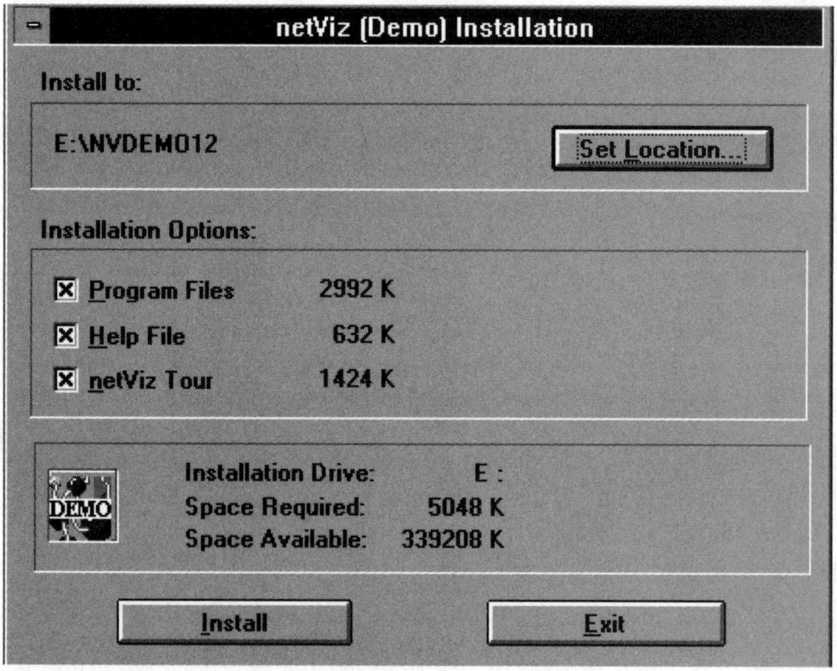

netViz (Demo) Installation		
Install to:		
E:\NVDEMO12		Set Location...
Installation Options:		
☒ Program Files	2992 K	
☒ Help File	632 K	
☒ netViz Tour	1424 K	
DEMO	Installation Drive:	E :
	Space Required:	5048 K
	Space Available:	339208 K
Install		Exit

Figure 6.32 You can choose to install the tour or the working model separately.

The installation process creates the Program Manager Group and icons, as the next figure (Figure 6.33) shows.

Figure 6.33 netViz installation creates the icon and group automatically.

To invoke netViz, click on an icon in the netViz group. If you want to try the program, click on the netViz ScreenCam Demo, then double-click on the netViz Demo. If you select the guided tour icon, you should see the automated screen and extra mouse cursor as shown in Figure 6.34.

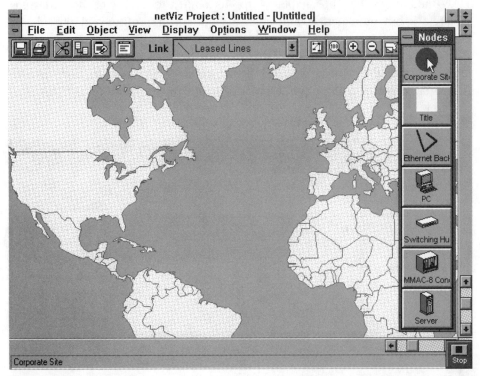

Figure 6.34 The netViz tour video.

When you want to stop the automated presentation, you will need to find the Lotus ScreenCam window and click on the red stop button. The other buttons are functional as well in the separate ScreenCam window, as Figure 6.35 shows.

Figure 6.35 End the netViz tour video by clicking on the end button.

The Lotus ScreenCam control panel has six features. The bar display indicates the relative elapsed portion of the movie. The four buttons stop the movie (temporarily), rewind the movie, fast forward it, and stop it altogether. The vertical slider increases the playback volume for the announcer's commentary.

LANBuild

LANBuild is profiled in Chapter 4 for its ability to verify network designs, generate a bill of materials and energy requirement profile, and provide the consummate services for network configuration. This is a shareware demonstration and a limited version of the full commercial application. You are allowed to pass the software demo along subject to copyright restriction. Installation is accomplished through Program Manager by running SETUP.EXE, as shown in Figure 6.36.

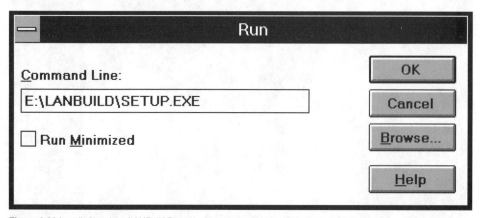

Figure 6.36 Installation of the LANBuild Demo is through the standard Windows setup process.

You can use the program to print or create screen shots of larger networks. The full installation requires about 1.5 MB. You have the option of installing to a directory of your choice, as Figure 6.37 shows.

The installation process creates the Program Manager Group and icons, as Figure 6.38 shows. To invoke LANBuild, click on the icon with the tools. You are limited to networks with 20 nodes.

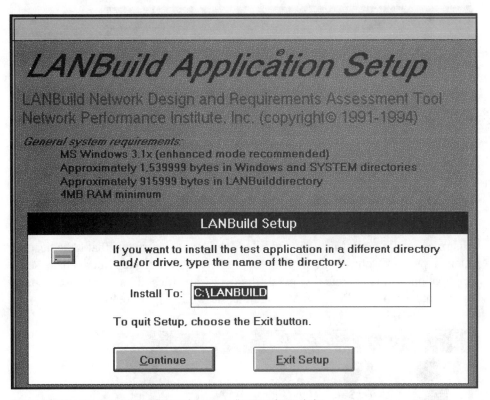

Figure 6.37 You can choose to install the software to a directory of your choice.

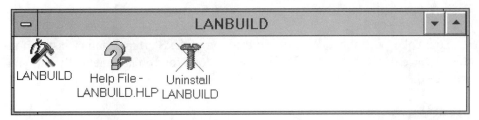

Figure 6.38 LANBuild installation creates the icon and group automatically.

Win, What, Where

Win, What, Where is profiled in Chapter 4. It tracks process and file access in Windows. It is a shareware demonstration application and must be registered with the current registration fee if you use it. You are allowed to pass the software demo along, subject to copyright restrictions. Installation is accomplished through Program Manager by running SETUP.EXE, as shown in Figure 6.39.

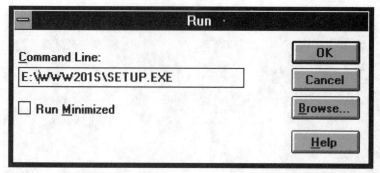

Figure 6.39 Installation of Win, What, Where is through the standard Windows process.

Specify a destination drive. Note that no other Visual Basic applications can be running during the application process. This includes NetSpecs, Disk Historian, or Program Manager and File Manager replacements that have become such a part of your desktop that you have since forgotten them. The setup application will expand and copy all the necessary files, do some housekeeping, and create a group in Program Manager, as shown in Figure 6.40.

Figure 6.40 Win, What, Where installation creates the icon and group automatically.

To invoke Win, What, Where, add the icon to the Startup group or install it to the RUN= entry in WIN.INI. This is usually done for you automatically and you will see the minimized task icon as shown in Figure 6.41.

Figure 6.41 The Win, What, Where minimized task.

If you want to view current operations, restore the minimized icon. If you want to view the more comprehensive databases, then double-click on one of the icons in the Program Manager group and you will see the tracking database screen as shown in Figure 6.42.

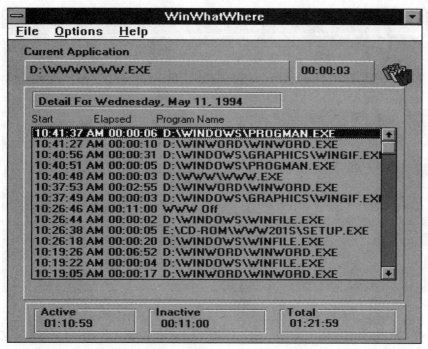

Figure 6.42 The Win, What, Where tracking screen.

Rummage

Rummage displays files with duplicate filenames. Rummage is a shareware application and must be registered with the current registration fee if you use it. You are allowed to pass the software along, subject to the stated copyright restrictions. Installation is accomplished through Program Manager by running INSTALL.EXE, as shown in Figure 6.43.

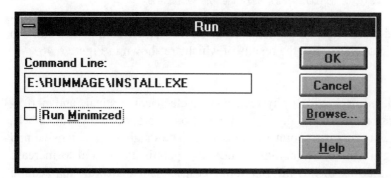

Figure 6.43 Installation of Rummage is through the standard Windows process. Note that the installation application is called INSTALL.EXE.

You will see an installation like the one shown on the next page (Figure 6.44).

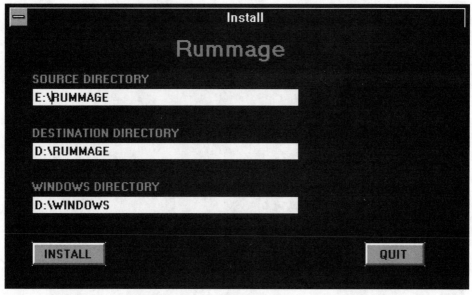

Figure 6.44 Installation of Rummage is quite simple and straightforward.

When complete you should have a monster icon in your Program Manager. If not, simply drag the RUMMAGE.EXE filename from File Manager into a Program Manager group of your choice, as shown in Figure 6.45.

Figure 6.45 Rummage icon is the monster, here installed in the accessories group.

When you double-click on the icon, click on the search command button to search for duplicate filenames, the results of which are shown in Figure 6.46.

NetSpecs

NetSpecs displays mechanical, physical, signal, electrical, optical, and performance information on many data communication protocols for both LANs, MANs, WANs, and point-to-point transmissions. The database is extensive and always growing to reflect the common and new specifications, and to increase the accuracy and utility of this information. NetSpecs is a shareware application and must be registered with the current registration fee if you use it. You are allowed to pass the software along, here also subject to the vendor's copyright restrictions. Installation is accomplished through Program Manager by running SETUP.EXE, as shown in Figure 6.47.

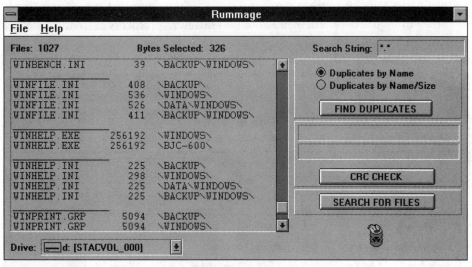

Figure 6.46 The Rummage report shows files with duplicate names with size and location.

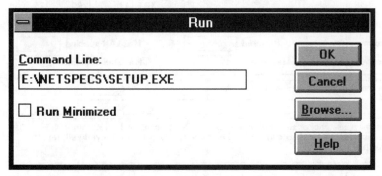

Figure 6.47 Installation of NetSpecs is through the standard Windows process.

You will be asked to specify a destination drive. Note that no other Visual Basic applications can be running during the application process. This includes Win, What, Where and Disk Historian. The setup application expands and copies the necessary files, does housekeeping, and creates a group in Program Manager, as shown in Figure 6.48. You can drag the NetSpecs icon into another folder and delete it in order to conserve folders or consolidate desktop space.

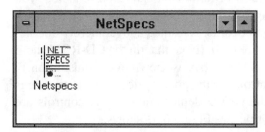

Figure 6.48 NetSpecs installation creates the icon and group automatically.

To invoke NetSpecs, double-click on the icon and you see the database screen as shown in Figure 6.49. There are about 70 different protocols currently listed in the database.

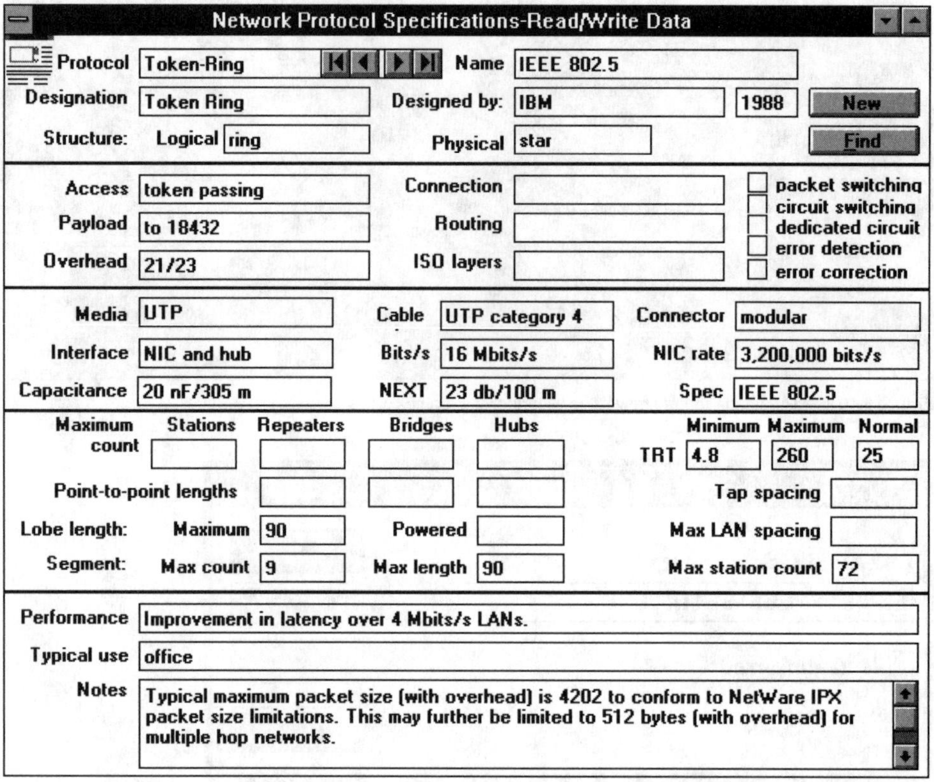

Figure 6.49 NetSpecs protocol database information screen.

ScreenCam movies

The CD-ROM has four Lotus ScreenCam movies in uncompressed format; one is in the NETVIZ directory. You can play them directly by clicking on the filename in File Manager. You do not have to load these files onto a hard drive (unless your system is very slow) as they work directly from the CD-ROM. The movies include sound so that if you have Windows Multimedia extensions enabled, you can listen to my running commentary about how to actually use these tools for designing, modeling, and blueprinting networks. As stated above, merely double-click on the filenames, as shown in Figure 6.50. (Note that on the CD-ROM these files are now in the root directory, not in \MOVIES as shown in the illustration.)

You should see the ScreenCam controls for the presentation once the movie is enabled, just as you described with the netViz demonstration. The controls are the same for each of the three presentations, as Figure 6.51 shows.

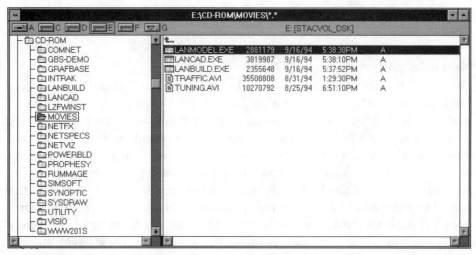

Figure 6.50 Click on the exectuble file to invoke the ScreenCam movie.

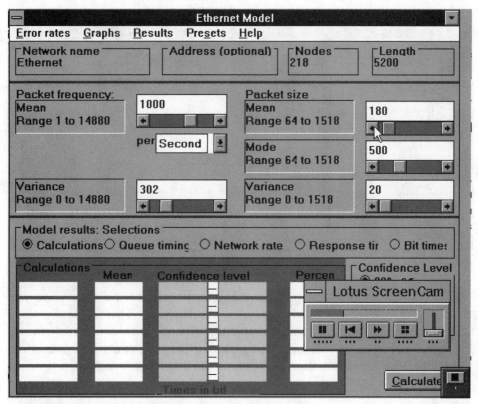

Figure 6.51 A Lotus ScreenCam movie as displayed from the screen.

The ScreenCam control panel has six features. The bar display indicates the elapsed portion of the movie. The four buttons stop the movie (temporarily), rewind the movie, fast forward it, and stop it altogether. The vertical slider increases the playback volume for my commentary.

Morph design movies

The CD-ROM has two morph movies in uncompressed format. You can play them directly by clicking on the filename in File Manager. You do not have to load these files onto a hard drive (unless your system is very slow) as they work directly from the CD-ROM. The movies include sound so that if you have Windows Multimedia extensions enabled, you can listen to my running commentary. The first movie shows how disparate types of highway traffic patterns which create similar jams actually require careful analysis to defuse. The second movie walks you through the optimization of premise wiring configurations. Invoke these movies just as did for the ScreenCam demonstrations, merely double-clicking on the filenames, as shown in Figure 6.52. (These have also been moved from \MOVIES to the root directory.)

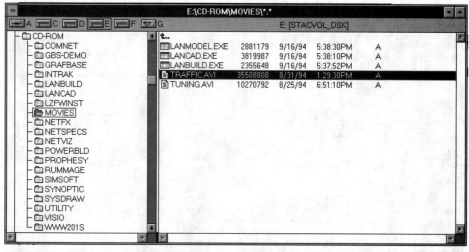

Figure 6.52 A dubbed morph sequence showing how various traffic patterns, loading, and composition create jams.

The many illustrations from Chapter 2 fit together and tell a story when you see each one transform into the next one in the sequence. This movie discusses time-of-day issues, traffic composition, timing, loading, and second-order effects. Figure 6.53 shows the transformation midpoint between two traffic scenarios. The blurriness in this illustration is a function of motion between frames.

Figure 6.53 A dubbed morph sequence showing how various traffic patterns, loading, and composition create jams.

The second movie shows how to transform the logical and physical design of several networks to simplify network maintenance, migrate to a structured premise wiring facility, and resolve design-induced performance bottlenecks. This movie is based upon figures from Chapter 3 and Chapter 5. Figure 6.54 shows a midpoint transformation between two similar network configurations.

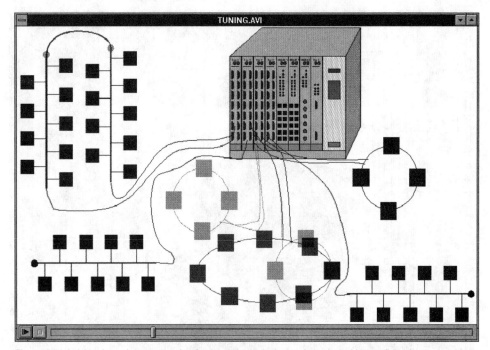

Figure 6.54 A dubbed morph sequence showing how various logical and physical wiring configurations can address and resolve enterprise network performance bottlenecks.

Sources

Bibliography and references

110 Alert, PC Power & Cooling, Inc., Carlsbad, CA.

AIX 3.2 RISC System/6000 Introduction to Performance Tuning, AIX National Technical Support Center, IBM Corporation, Southlake, TX.

AIX 3.2 RISC System/6000 Advanced Performance Analysis and System Tuning, AIX National Technical Support Center, IBM Corporation, Southlake, TX.

AIX 3.2 RISC System/6000 Performance Monitoring and Tuning Guide, IBM Corporation, Austin, TX.

AIX Performance Monitoring and Tuning, Virgil Albaugh, *AIXpert Journal*, Winter 1991.

AllMicro, 1994, *LAN Designer,* AllMicro, Clearwater, FL.

Abstraction Software, 1994, *Prophesy*, Abstraction Software, CO.

American HyTech Corporation, 1994, *NetGuru*, American HyTech Corporation, Pittsburgh, PA.

AMIDiag, American Megatrends, Norcross, GA.

Aquino, Michael, SynOptics Communications, Inc., Atlanta, GA.

B&B Electronics Mfg, 1994, *Green Keeper*, B&B Electronics Mfg., Ottawa, IL.

Bachman Information Systems, 1993, *The Performance Perspective; Designing Performance Into Your Applications,* Bachman Information Systems, MA.

Banyan Systems, 1992, *Monitoring and Optimizing Vines Performance*, Westborough, Banyan Systems, MA.

Barlow, Brian, Future Systems Solutions, Angola, IN.

Basic Systems, 1994, *Win, What, Where,* Time and Usage Monitor for Windows, Basic Systems, Inc., Kennewick, WA 99336.

Beshore, Ed, September 27, 1993, *NFS Benchmark: Nice, But Not Enough*, Open Systems Today, CMP Publications, NY.

Bierbaum, Neal, Make Systems, Inc., Mountain View, CA.

Birchall, Graeme, 1994, *DB2/2 Access Paths and SQL Tuning*, CompuServe 73540,1566.

Birchall, Graeme, 1994, *Tuning SQL Statements*, CompuServe 73540,1566.

Blitz 'n' Software, 1994, *CD-Blitz*, Blitz 'n' Software, Inc., San Jose, CA.

Bodkin, Richard, Account Representative, Magnetic Software, Mount Airy, NC.

Box, Mark, VP Sales and Marketing, Pure Software, Sunnyvale, CA.

Brown, Matt, Vice President-Sales, Microsystems Engineering Company, Lombard, IL.

Burress, Randy, Public Relations Specialist, Intel Corporation, Hillsboro, CA.

Business Applications Performance Corporation, 1991, *Research Methodology, Computer Performance Benchmark Survey,* Business Applications Performance Corporation (BAPCO.), CA.

Chappell, Laura, 1993, *NetWare LAN Analysis*, Novell Press (Sybex), San Jose, CA. ISBN 0-7821-1143-2.

Chediak, Juan Carlos, cisco Systems, Miami Lakes, FL.

Clear & Simple, 1992, *Performance 2.1, A Tuning Kit for OS/2*, Clear & Simple, Inc., West Simsbury, CT.

Conklin, Dave, IBM Independent Vendor League, Greenwich, CT.

Conklin, Dick, Editor, *OS/2 Developer Magazine*, IBM Corporation, Boca Raton, FL.

Day, Michael, November 18, 1991, *Tuning NetWare File Server Performance*, *LAN Times*, McGraw-Hill, CA.

Derfler, Frank J. Jr., and Schireson, Max, *Maximizing the Performance of Your Server*, *PC Magazine*, October 26, 1993.

Doc-to-Help, version 1.1, WexTech Systems, Inc., New York, NY.

Dodson, Doug, President, PC Power & Cooling, Inc., Carlsbad, CA.

Diskeeper/Plus, Executive Software, Glendale, CA.

DrivePro, Micro House, Boulder, CO.

Eaken, Kenneth, Vice President, Marketing, MapLinx Corporation, Plano, TX.

Easton, Richard, President, Basic Systems, Inc., Kennewick, WA.

Editors, May 1991, *Tweaking the Tail of the Beast, Windows Tech Journal.*

Editors, June 1992, *The Need for Speed, MacWorld*, CMP Publications, NY.

Editors, November 2, 1992, *CPU Upgrade Easiest Route to Faster Speeds, MacWeek*, NY, NY.

Editors, 1993, *Performance Enhancers, MacUser.* Annual Review, NY, NY.

Editors, April 1993, *Supercharging Your Mac, MacUser*, NY, NY.

Epson Wingine, Epson America, Carlsbad, CA.

Ernst, Ted and Schwee, David, February 1993, *Tuning NCP To Improve SNA Network Performance,* IBM Corporation Technical Support, TX.

Fraga, Jose, Systems Engineer, 3Com Corporation, Boca Raton, FL.

Frank, Maurice, April/May 1994, *Database Design Dilemmas, Programmer's Journal,* R&D Publications, KS.

Future Systems Solutions, 1992, Speedcache+, Future Systems Solutions, Inc., Bluffton, IN.

Glass, Brett, March 1994, *CD-ROM Drive Acceleration Programs: Modest Speedups, No Miracles,* Multimedia World, Carlsbad, CA.

GMS, Probots, Inc., Northhampton, MA.

Goguen, Gerry, Product Manager, Open/Access, Wang Laboratories, Lowell, MA.

Gould, David, VP, Business Development and Marketing, Wang Laboratories, Lowell, MA.

Green, Emily and Collins, Stephen, August 9, 1993, *Is Your Network About to Collapse?, LAN Times,* p. 43, McGraw-Hill, CA.

Haight, Timothy, *Internetworks by Great Leaps and Measured Steps, Network Computing*, December 1, 1992, CMP Publications, NY.

Harrison, Peter G., and Patel, Naresh M., 1992, *Performance Modelling of Communication Networks and Computer Architectures*, Addison-Wesley Publishing, MA. ISBN 0-201-54419-9.

HawkNet, 1994, *NetTune*, Director of Software Engineering, HawkNet, Inc., Carlsbad, CA.

Heise, Russell A., September/October 1992, *Performance Tuning: A Continuing Series, /AIXtra*, IBM's Magazine for AIX Professionals, IBM, Boca Raton, FL.

Heise, Russell A., Performance Team Leader, AIX Performance Consultant, IBM AIX National Technical Support Center, IBM Corporation, Roanoke, TX.

Hildebrand, J.D., Volume 9, Number 9, *The Well-Tempered Network, Unix Review*, pp. 29–31.

Hughes, John, Technical Support, Ingres Corporation, The ASK Group, Alameda, CA.

IBM Corporation, 1991, *Local Area Network Server Version 2.0, Information and Planning Guide,* IBM, White Plains, NY.

IBM Corporation, 1991, *OS/2 V2.1 Redbook*, IBM, IBM Corporation, International Technical Support Center, Boca Raton, FL.

IBM Corporation, 1992, *IBM OS/2 LAN Server Version 3.0 Network Administrator Reference, Volume 2: Performance Tuning,* IBM Corporation, Austin, TX.

IBM Corporation, 1992, *LAN Host Gateways Function and Selection Guide*, ZZ81-0299-00, IBM Corporation, Armonk, NY.

IBM Corporation, 1993, *OS/2 2.1 Performance Tuning*, IBM International Technical Support Center, Boca Raton, FL.

IBM Corporation, 1993, *OS/2 2.1 Performance Tuning for End Users*, IBM, OS/2 System Performance Department, Boca Raton, FL.

IBM Corporation, 1993, *TPC Benchmark: A Full Disclosure Report*, IBM Corporation, TX.

Ingres Corporation, 1993, *DBMS Optimizations for Transaction Processing*, The ASK Group, Alameda, CA.

Intel Corporation, 1993, *Product Overview,* Intel Corporation, IL.

Jacobson, Janice, Marketing Manager, PC Power & Cooling, Inc., Carlsbad, CA.

Jazwinski, Andy, President, Network Performance Corporation, Dunkirk, MD.

Kantor, Marc, Director of Marketing, Solid Oak Software Inc., Santa Barbara, CA.

Khalil, K.M., and Luc, Q.V. and Wilson, D.V., 1990, *LAN Traffic Analysis and Workload Characterization,* Bell Communications Research, IEEE 0742-1303/90/0000/0112, NY.

Kolodziej, Stan, February 1994, *Rightsizing Rewrites Rules for Managing Performance*, *Software Magazine*, MA.

LaserMaster Corporation, 1994, WinJet, LaserMaster Corporation, Eden Prairie, MN.

Lebaron, Chris, Product Manager, CACI Products Company, La Jolla, CA.

Leland, Will, and Willinger, Walter, and Taqqu, and Murad and Wilson, Daniel, October 1993, *On the Self-Similar Nature of Ethernet Traffic*, ACM SIGCOMM 1993, The SIGCOMM Quarterly Publication, Volume 23, Number 4. IEEE, NY.

Levitt, Jason, January 20, 1992, *Better Benchmarks Are Brewing*, *Unix Today*, CMP Publications, NY.

Lewallen, Dale, 1993, *This Old PC*, Ziff-Davis Press, NY.

Lewis, Elizabeth, March/April 1993, *Performance Tuning: Theory and Practice*, */AIXtra*, IBM's Magazine for AIX Professionals, Boca Raton, IBM, FL.

Pietrek, Matt, *Liposuction Your Corpulent Executables and Remove Excess Fat*, *Microsoft Systems Journal*, July 1993.

Lough, Mike, Caere Corporation, Los Gatos, CA.

Loukides, Mike, 1991, *System Performance Tuning*, O'Reilly Books, CA. ISBN 0-937175-60-9.

MapLinx, MapLinx Corporation, Plano, TX.

Matrox Electronic Systems, Ltd., Dorval, Quebec, Canada.

Marion, Larry, April 1994, *Keeping Your Balance: Managing a Multivendor Client/Server Shop, Beyond Computing*, NYT Custom Publishing, NY.

McDonald, Kathleen S., Microsystems Engineering Company, Lombard, IL.

Microsoft Corporation, 1992, *The Windows Resource Kit 3.1,* Microsoft Corporation, WA.

Microsoft Corporation, 1993, *Microsoft Guide to Optimizing Windows*, Dan Gookin, Microsoft Press, Redmond, WA.

Microsoft Corporation, 1993, *TPC_B Results for Microsoft SQL Server on Windows NT Running on Compaq ProLiant Systems*, Microsoft Corporation, WA.

Microsoft Corporation, 1994, *Optimizing the Microsoft LAN Manager TCP/IP Protocol,* Microsoft, WA.

Microsoft Corporation, 1994, *The Windows for Workgroups Resource Kit 3.12 Addendum,* Microsoft Corporation, WA.

Microsystems Engineering Company, 1994, *SysDraw,* Microsystems Engineering Company, Lombard, IL.

Milburn, Brian, Solid Oak Software Inc., Santa Barbara, CA.

Miller, Mark, March 21, 1994, *Life on the Fast LAN,* Mark Miller, *Network World,* pp. 41–49, International Data Group, MA.

Mills, Brendon, Senior Product Marketing Manager, Networth, TX.

Minoli, Daniel, January 10, 1994, *Designing Scalable Networks*, *Network World*, p. 17, International Data Group, MA.

Naples, Anthony, Wellfleet Communications, Inc., Boca Raton, FL.

Nemzow, Martin, 1993, *LAN Performance Optimization*, McGraw-Hill, NY. ISBN 0-8306-4277-3.

Nemzow, Martin, 1994, *Computer Performance Optimization*, McGraw-Hill, NY. ISBN 0-07-911689-2.

Neshamkin, Paul, Technical Support Coordinator, WexTech Systems, Inc., New York, NY.

Network Performance Institute, 1993, *NetSpecs,* Network Performance Institute, Miami Beach, FL.

Network Performance Institute, 1992, *LAN•CAD,* Network Performance Institute, Miami Beach, FL.

Network Performance Institute, 1990, *LANBuild,* Network Performance Institute, Miami Beach, FL.

Neuenschwander, Bill, LasterMaster Corporation, Eden Prairie, MN.

Ostrow, Gail, IBM Independent Vendor League, Greenwich, CT.

Pacific CommWare, *TurboCom/2,* Pacific CommWare, Ashland, OR.

Parr, Diana, Product Marketing, Ingres Corporation, The ASK Group, Alameda, CA.

Pendergrass, Barbara, Marketing Product Manager, SAS Institute, Cary, NC.

Peterson, John, Business Applications Performance Corporation (BAPCO), Santa Clara, CA.

Pietrek, Matt, *Liposuction Your Corpulent Executables and Remove Excess Fat, Microsoft Systems Journal*, July 1993.

Pittman, Robin, Marketing Specialist, Epson America, Carlsbad, CA.

Philip Blachier, 1994, *PowerBuilder Design Tips, CompuServe, Powersoft Forum,* 72632,2350.

Price, Michael C., Director of Software Engineering, HawkNet, Inc., Carlsbad, CA.

Probots, Inc, 1993, GMS, Probots, Inc., Northhampton, MA.

Proisise, Jeff, *Maximizing Memory Under DOS 6.0, PC Magazine*, August 1993, NY, NY.

Quyen Systems, 1994, *netViz 1.1,* Quyen Systems, Rockville, MD.

RenaSonce Group, 1992, Skylight, RenaSonce Group, Inc., San Diego, CA.

Richard, Kevin and Johnson, Eric, Volume 9, Number 9, *X Performance, UNIX Review*.

Roarabaugh, Virginia, Advisory Programmer in OS/2 Systems, Performance, IBM, International Technical Support Center, Boca Raton, FL.

Rongley, Eric, Blitz'n'Software, Inc. San Jose, CA.

Saettler, Mark, 1994, *Graphics Design and Optimization,* Multimedia Technical Support, Microsoft Corporation, WA.

Salemi, Joe, September 28, 1993, *Tuning DOS and WIN-OS/2 Sessions under OS/2 2.x, PC Magazine,* Ziff-Davis, NY.

Salerno, Richard, cisco Systems, Atlanta, GA.

Scherer, Pat, Fall 1992, *OS/2 Productivity through Multitasking*, IBM OS/2 Developer, IBM, TX.

Scherer, Pat, Industrial Engineer, IBM Personal System Line of Business, Austin, TX.

Sharma, Ranjana, President, Network Dimensions, Inc., San Jose, CA.

Shilling, Dale, SynOptics Communications, Inc., Hollywood, FL.

Shulman, Richard, Laserprint, Pompano Beach, FL.

SimSoft Inc., 1994, *SimSoft,*SimSoft Inc., Brooking, OR.

Smith, Mark, January 1992, *Analyzing Performance Problems on Internets, Networking Management,* CMP Publications, NY.

Solid Oak Software Inc., 1993, *PC Sentry*, System Activity Monitoring, Logging, and Security Software for PCs & Networks, Solid Oak Software Inc., Santa Barbara, CA.

Solid Oak Software, 1994, *Disk Historian*, File Activity Usage Monitoring Software for IBM PCs, Compatible Computers, and peer-to-peer and server networks, Solid Oak Software Inc., Santa Barbara, CA.

Stevens, Larry, March 1, 1994, *Hub/Router Combos: Don't Overwork Them, Datamation*, BPG, MA.

Stickle, Don, May 1994, *DB2 System Tuning, Enterprise Systems Journal,* pp. 45–50, Cardinal Business Media, TX.

Strehlo, Kevin, Editor-in-Chief, March 1, 1994, *Run It Over Twice, Then Benchmark It, Datamation*, BPG, MA.

Sun Microsystems, December 1990, *Network and File Servers: A Performance Tuning Guide*, Sun Microsystems, CA.

Sun Microsystems, 1991, *Sun Performance Tuning Overview,* Sun Microsystems, CA.

Sun Microsystems, 1992, *Management Solutions for Commercial Open Systems; Guide to Vendors and Products*, Sun Microsystems, Mountain View, CA 94043.

SuperSet Software Corp., 1992, WinGIF, SuperSet Software Corp., Provo, UT.

TechPool, 1991, *Atmospheres Background Systems*, TechPool, Cleveland, OH.

Tuma, Wade, President, Disk Emulation Systems, Inc., Santa Clara, CA.

Wade, Virginia, Technical Support Specialist, WexTech Systems, Inc., NY, NY.

Wang Laboratories, 1994, *OPEN/Image,* Wang Laboratories, Lowell, MA.

Wann, Robert, Vice President, Blitz 'n' Software, Inc., San Jose, CA.

Ward, Bonny, Vice President, Advanced Visual Data, Inc., Waltham, MA.

Warmbrod, Bruce, Intel Corporation, Norcross, GA.

Weston, Rusty, June 1993, *Optimizing your Mainframe,* Corporate Computing, M&T Publishing, CA.

Whitehair, Bob, President, Probots, Inc., Northhampton, MA.

Whitfield, L. Kenneth, Advisory Systems Analyst, IBM, Austin, TX.

Wilson, Brian, Eastern Area Systems Engineer, Sun Microsystems, Mountain View, CA 94043.

Wilt, Richard, Director of Marketing, American HyTech Corporation, Pittsburgh, PA.

Index

.INI, 234, 240
/ETC/HOSTS, 214
3Com, 221, 223, 245
10Base-T, 52, 77, 177, 221
10Base5, 77, 78
100Base-T, 52, 77, 81, 221
100Base-X, 81
100BaseVG-AnyLAN, 77, 81, 186, 211, 221
110 Alert, 232-233

A

accidental latency, 28-29
activity tracking, 124
address resolution, 231
ADM, 194
Advanced Visual Data, 153
AIX, 4, 120
Alantec, 221
AnyNet, 235
API-MPAPI, 137
Apple Macintosh, 124, 264
  System 6.x, 4
  System 7, 4
AppleTalk, 235
application code, optimizing, 241-244
APPN, 67, 245
ARCNET, 74, 77, 88, 271
ASIC, 224
ATM, 33, 61, 70, 88, 104, 132, 182, 183, 191,
    192, 194, 199, 211, 212, 214, 217, 225,
    231, 237
ATM switch, 70
AURP, 218
AutoCAD, 105
AutoSys, 123

B

backbones, 86-87, 185
  collapsing, 209-210
backplanes, 84, 87, 185, 210
  vs. hubs, 52

backward explicit congestion notification
    (BECN), 231
Balance, 262
Balance NLM, 223
bandwidth, 25-26, 40, 41, 42, 48, 50, 59, 61-
    66, 70, 72, 73, 75, 104, 162-163, 167, 226
  ATM, 33
  channel, 145, 168
  dedicated, 47-48
  I/O, 65
  increasing transmission, 210-213
  limitation of WAN, 46
  memory, 65-66
  mobile/wireless, 65
  on demand, 214 power, 230-231
  utilizing, 61-65
bandwidth congestion, 19
  segment, 19
BAPCo, 93, 141
BAR, 119
BatchScheduler, 123
BECN, 231
benchmarks, 92-93
  applying, 143
  disadvantages, 140
  types of, 140-143
bidirectional line-switched ring (BLSR), 193,
    195
BIND, 123
BIOS, 75
Bitfax, 179
blocking, 84
BLSR, 193, 195
blueprinting, 103-109
BMP, 238
BONeS, 148, 162
bottlenecks, 13, 15, 18, 19, 24, 30, 31, 33, 55-
    59, 167, 179, 188, 215, 223, 226
  definition, 55-59
  determining causes of, 92-93
  disconnected devices and, 91-92
  items contained within, 58

bottlenecks *continued*
  reasons for, 59-61
boundary routing, 179-180, 275
Bradner, Scott, 141
breakpoint, 66
bridges, 76, 79-80, 215-216
  vs. routers, 155-160
BTU, 111
buffered network I/O, 222
buffer overrun, 21
buffer size, router, 161-162
bursts, 29-33

C

C, 66, 128, 130, 247
C++, 66, 128, 130, 243, 247
cable certification, 131-132
Cabletron, 41, 221
Cabletron MMAC, 229
Cabletron Spectrum, 110
cache configuration, 238-241
CAD, 65
CAD/CAM, 75
Cairo, 247
capacity
  channel, 145
  network, 145
  planning, 155
CBR, 231CBS, 123
cc:Mail, 275
CD-Blitz, 240, 269
CD-ROM, 98, 106, 121, 124, 132, 182, 240-
    241, 251-258, 269, 279-312
  demonstrations, 287-312
  features, 279
  file contents, 280
  handling, 279
  help format, 277-283
  installing, 281
  MPC-compliant reference, 3, 8
  multimedia viewer format, 283-287
  on-line hypertext, 281-287
  traffic congestion example, 13
CDDI, 132
channel bandwidth, 145, 168
channel capacity, 25
Chicago, 247
CICS, 63, 240
Cisco, 245
clients, 76, 268-272
client/server transaction processing, 248-266
CMG, 136
CMOS, 75, 110
CMS, 240
COBOL, 253
COMMIT, 123
common object request broker architecture
  (Corba), 137
common open software environment
  (COSE), 137

communications
  data, 1, 34-39, 37-39
  modem, 40, 94-95, 214, 275-276
  remote, 94-95
COMNET, 162-163
COMNET III, 163, 293-294
component software, 124
compression, 118
CompuServe, 275
computing, mobile, 92
computing platforms, 4
CONCURRENCY, 123
configuration management, 94
control block sampling (CBS), 123
Corba, 137
COSE, 136, 137
costs, 47, 112, 201-205
CPM, 57
CPU, 65, 75, 93, 95, 119-120, 266
critical path, 56-57, 99
CRONTAB, 120, 121
crosstalk, 77
CSU/DSU, 246, 275
customer subscriber identification number
  (CSID), 203

D

Da Vinci, 275
DACS, 194
dark fiber, 213
data communications, 1, 34-39
  leaks and losses, 37-38
    mechanical infrastructures, 35-37
    pressure and volumes, 38-39
data compression, 118
data conversion, 243
data encapsulation, 52, 81
data processing, 52
data sharing, 47-48, 74-75
data translation, 52
Daytona, 247
DB2, 7, 123, 256-257
DBAnalyzer, 122
dBase, 93
DBMS, 122
DBTune, 122
DCE, 137
DD1, 123
DEADLOCK, 123
DECnet, 118, 199, 218, 235
delay, 27-28
deleting, unwanted files, 125-126
demand priority, definition, 211
desktops, optimizing, 234
devices, tracking disconnected, 241
Dhrystones, 141
DID, 203
direct-inward dial (DID), 203
disaster recovery plan, 112-123
discrete modeling, 147-148

Disk Historian, 125-126
Distributed Computing Environment (DCE), 137
distributed management environment (DME), 137
distributed systems, rightsizing and control of, 93-97
distributed transaction processing (DTP), 123
DLSw, 245
DME, 137
DNS, 206
domain name system (DNS), 206
DOS (*see* MS-DOS; operating systems; PC-DOS)
DSL, 88
DSU/CSU, 63, 235
DTMF, 203
DTP, 123
dual-tone multifrequency (DTMF), 203
duplex signaling, 40
dynamic switching (*see* port switching)

E

E-mail, 199, 202ECC, 268
Eclipse FAX, 179
EGP, 218
EIA/TIA, 179, 180, 186, 196, 211
EISA, 75
EMS, 271
emulation modeling, 153-154
encapsulation, definition, 81
end nodes, 76
end-to-end latency, 68-69
energy usage, 231-232
engagement of capacity, 61
ENS, 7
enterprise hub, 81
enterprise network
  components, 57-58
  contents of, 11
  definition, 12
  description/overview, 12-24
  designing, 195-201
  global view, 36
  infrastructure, 23-24, 176-177
  integration, 198-200
  modeling, 145-172
  optimization methods, 177-179
  optimizing with design, 179-195
  purpose, 196
  redundancy, 197-198
  scalability, 200-201
  schematic, 36
  security, 198
  wiring infrastructure, 196
enterprise router, 81
EPA Energy Star program, 231
equipment, purchasing used, 204
error correction code (ECC), 268
errors, wiring, 61

ES9000, 235
Ethernet, 16, 36, 40, 47, 61, 74, 77, 88, 155, 221
Ethernet 802.3, 211
extended memory (EMS), 271
extrapolation modeling, 146

F

failure rate, 165
FAServer, 267
fax, 97, 179, 202-203
fax imaging, 99
faxPAD, 204
FDDI, 16, 36, 40, 46, 47, 52, 61, 74, 77, 88, 132, 155, 177, 182, 183, 185, 210, 217, 221, 225
FECN, 231
FEP, 219
Fibre Channel, 206, 212
file transfer protocol (FTP), 91
files, deleting, 125-126
FileState, 125
finances (*see* costs)
firewalls, 45, 81
flow control mechanism, 71
FOIRL, 211
fonts, 234
forward explicit congestion notification (FECN), 231
FoxPro, 247
fragmentation, 102, 268
frame relay, 104
frame size, 33
FSCK, 121
FTP, 90, 91, 238
full-duplex transmission, 229-230

G

gateways, 76, 83-84, 198, 214-215
  SNA performance optimization, 257-259
GDI printing technology, 119
GigaSwitch, 226
GIS, 106
GMS, 160, 161
GMS 2.0, 159
GrafBase, 106, 288-289
graphical user interface (GUI), 270
Green Keeper, 232-233
gridlock, 18-19, 22, 81
GUI, 270

H

handicapped wait, 238
hardware, recycled equipment, 204-205
Hawknet, 145
hierarchical storage management (HSM), 75, 126
  levels of, 207
high memory (HMA), 271
HINT, 141

HLDC, 7
HMA, 271
HP, 127
HP LaserJet II, 265
HSM, 75, 126, 137, 207
hubs, 40, 41, 76, 78-79, 84, 185, 221-222,
   226, 229
   enterprise, 81
   optimizing, 275-276
   vs. backplanes, 52, 51
HVAC, 231

I
I/O bandwidth, 65
IBM, 127
IBM Power PC, 247
IDE, 75
IDNX, 153
IEEE 801.10d, 192
IEEE 802.3, 88
IEEE 802.9, 212, 223
IEEE 802.9a, 230
IETF, 136, 220
IFI, 123
IGRP, 218
imaging, 98
   fax, 99
   optimization, 261-262
IMF, 138
impedance mismatch, 77
information storage management, 94
instrumentation facility interface (IFI), 123
integration, 52-53
Intel 486, 247
Intel 8086, 64
interconnectivity devices (see specific types of)
interexchange carrier (IXC), 97
intermediate nodes, 76
Internet, 275
Internet Engineering Task Force (IETF), 136
internetwork management forum (IMF), 138
inventory tracking, 110-111
inverse multiplexing, 182
IOSTAT, 120, 121, 122
IP, 81, 218
IPX, 218
IPX/SPX, 37, 88, 90, 91, 117, 218, 237
IRGP, 149
IRMAlink, 198
IS-IS, 218
ISA, 75
ISDN, 191, 209, 214, 237, 275
ISO, 214
isoENET, 212, 223
IXC, 97

J
jitter, 77

K
Kalpana, 221

L
LADDIS, 142

LAN (see local area network)
LAN Designer, 111
LAN Manager, 4, 7, 90, 115, 118, 276
LAN Server, 4, 7, 90, 115, 238, 276
LAN-CAD, 105-106, 298-299
LANalyzer, 97, 132, 134, 144, 295-296
LANBuild, 109, 111, 113-114, 233, 302-303
LANCAD, 109
LANDesk, 71, 133-134
LANModel, 109, 151, 152, 168-170
   transaction modeling, 171
LANNET MultiNet, 84
Lannet, 221
LANStreamer, 229
LANtastic, 276
large internet packet (LIP), 238
LaserMaster, 75, 119, 265
latency, 27-28, 45-46, 59, 66-71, 72, 145, 162-
   163, 167
   accidental, 28-29
   comparisons, 68
   decreasing transmission, 208-209
   end-to-end, 68-69
   measuring, 63, 67
   mobile/wireless, 71
   packet, 66
LEC, 97
Leverage, 118
LEX, 130
Lexmark, 265
link simulator, 105
LIP, 238
LLC, 81, 245
LLC2, 89
load, 45-46, 48, 71-73
Load Balancer, 123
local area network (LAN), 40-45, 60, 83, 86,
   89, 93, 102, 104, 130, 131, 132, 134, 145,
   154, 162, 177, 183, 185, 199, 211, 214,
   217, 246
   bandwidth, 64
   connecting to WAN, 235
   firewalls, 45
   modeling, 48-49
   operations, 44
local exchange carrier (LEC), 97
logic operators, 243
Lotus 1-2-3, 93

M
MAC, 81
MAC-layer, 90, 224-225
MAC-level, 79, 160, 171
Madge, 221
mail handling service (MHS), 275
Make Systems Analyzer, 191
MAN, 45-46
management, update, 130-131
management facilities, 94-97
management information base (MIB), 116
MapExpert, 103

MapLinx, 103
matrix switch, 84
MAU, 77, 78
MCI Mail, 275
mean time between failure (MTBF), 165
media, sharing, 47-48, 74-75
MEMMAKER, 117
metropolitan area network (MAN), 45-46
MHS, 7, 275
MIB, 116, 228
MIB objects, 135
Micom, 245
microsegmentation, definition, 207-208
Microsoft, 127
Microsoft Mail, 275
Microsoft Office, 240
MIPS, 141
mobile computing, 92
modeling, 48-49, 145-172
    discrete, 147-148
    emulation, 153-154
    extrapolation, 146
    simulation, 147-149
    state-change, 147-148
    statistical, 150-153
    transaction, 171
models, 101
modem communications, 40, 94-95, 214, 275-276
MONITOR, 97, 121, 122, 134, 145
Morph, 310-312
Motorola 6502, 64
MQUIPS, 141
MS-DOS, 4, 110, 124, 240
MS Windows, 7
MSAU, 77, 78
MTBF, 165
MultiNet, 226
multiport router, 81-83
multiprotocol stacks, 90-91
    optimizing with, 235-236
multitasking, 247-248
multithreading, 247-248
MVS, 134

N
naming, 261
NASI, 276
NCP, 88, 118
NCSI, 276
NDIS, 236
NDS, 206
NET F/X, 289-290
NetBEUI, 155, 218, 236
NetBIOS, 89, 236, 245
NetGuru, 109-110, 148
NetMake, 123
NetShare, 123
NetSpecs, 306-308
NETSTAT, 63, 120

NetTune, 115, 145
NetView, 110
netViz, 35, 109, 299-302
NetWare, 4, 7, 39, 90, 110, 115, 118, 121,
    133, 139, 145, 199, 235, 240, 276
NetWare Core Protocol (NCP), 88, 118
NetWare directory services (NDS), 206
NetWare DOS Requester, 271-272
NetWare Multiprotocol Router, 91
network enterprise, aspects of, 24-34
network information service (NIS), 206
network switch, 84
networks and networking
    buffered I/O, 222
    capacity, 145
    data communications (see data communi-
        cations)
    distinction between different, 12
    enterprise (see enterprise network)
    LAN (see local area network)
    MAN, 45-46
    multihop, 181
    optimizing, 3-4, 175-277
    panic, 28-29
    protection, 45
    reliability/survivability, 164-167
    structured, 176
    traffic (see traffic)
    WAN (see wide area network)
NFS, 7, 90, 115, 120, 121, 132, 139, 140, 199
NFSWATCH, 63, 121
NIC, 75, 77, 121, 214-215, 223, 266
NIS, 206
NLM, 262, 267, 268
NLSP, 218, 234
NMS, 110
nodes, 76
NT Advanced Server, 4
NTAS, 7, 39, 115, 118, 121, 218, 240, 247
NuBus, 233

O
object management group (OMG), 137
object-oriented programming (OOP), 128,
    242-243
object request broker (ORB), 137
OCR, 262
ODI, 236
ODS, 41
OEM, 140
OLTP, 142
OMG, 136, 137
on-line documents, 97
on-line transaction processing (OLTP), 142
OOP, 128, 242-243
Open Software Foundation (OSF), 137
OpenView, 110, 119
operating systems, 4, 7
    optimizing, 236-241
operations management, 94

optimization, 175-277
  application code, 241-244
  definition, 4
  desktops, 234
  financial, 201-205
  generalized techniques, 195
  imaging, 261-262
  methods of, 177-179
  multithreading/multitasking, 247-248
  operating systems, 236-231
  purpose of, 7-8
  technical solutions, 205-231
  transaction processing, 246-247
  with design, 179-195
Oracle, 122
ORB, 137
OS/2, 4, 7, 124, 236, 247
OS/400, 115
OSF, 136, 137
OSI, 80, 235
OSI CMIP, 136
OSPF, 149, 218

P

pacing, 70
packet latency, 66
packet processing time, 67
packets, simulating size of, 167-171
packet switch, 42, 47, 224
packet transfer time, 67
PAGESET, 123
parts, 111
Pascal, 128
path protection switching (PPS), 191
Pathworks, 118, 199
PBX, 212
PBX lines, 104
PBX switch, 38, 42
PC DOS, 4, 110, 124, 240
PC LAN, 4
PC Sentry, 125
PC Server, 7
PCI, 75
peaks, 29-33
Pentium, 75, 247
PERFMETER, 122
PERFMON, 122
PERFORM2, 144, 230
PERFORM3, 144, 221, 230
Performance 2.1, 115
performance tools, 102-103
PERT, 57, 99
phased-locked loop (PLL), 77
PID, 120
Pinpoint, 128
pipelining, 223
PLL, 77
point-of-presence (POP), 212
POP, 212
port switching, 229
  definition, 221

PostScript, 265
power bandwidth, 230-231

power consumption, 231-232
PowerBuilder, 128, 243, 247, 255-256
PowerHub, 226
PowerOpen, 4
PPP, 155
PPS, 191
printer definition language (PDL), 264-265
printers, configuring, 262-265
private virtual circuit (PVC), 198
process flows, 107-110
process identification description (PID), 120
profilers, 127-130
Prophesy!, 157-159, 166, 297
protocol overhead, 26-27
protocols (see also specific types of)
  analyzing, 132-139
  differences between, 88-90
  file transfer, 91
  multiple stacks, 237-238
  multiprotocol stacks, 90-91, 235-236
  network, 90-91
  synchronous, 245-246
protocol stacks, 90
PS, 120, 122
PSSTAT, 120, 122
PSTAT, 122
PVC, 48, 198, 201, 214

Q

QANDA, 153
Quartet-signaling Ethernet, 88
QUEMM OPTIMIZER, 117
query govenor, 241
queueing system, 156
QUIPS, 141

R

RAD, 242
RAID, 167, 171, 239, 268
RAM, 97, 118, 239-240, 268
rapid application development (RAD), 242
RCP, 238
RCPSPY, 63
remote communications, 94-95
repeaters, 76, 78, 214
replication services, 259-260
Retix, 41
ringer circuits, 77
RIP, 90, 91, 218, 234
RISC, 224
RMON, 22, 97, 110, 135, 227
RoboMon, 118
route, 26
router convergence time, definition, 82
routers and routing, 41, 52, 76, 80-81, 217-218
  algorithms, 218-220
  boundary, 179-180, 275
  buffer size, 161-162
  enterprise, 81
  multiport, 81-83
  vs. bridges, 155-160

vs. switches, 160-161
RPC, 120
RTMP, 218
Rummage, 126, 305-306

S

SAP, 91, 234
SAR, 120, 122
SAS/CPE, 134
scalability, 49-50
scanners and scanning, 131-132, 262
SCO UNIX, 121
ScreenCam, 308-310
SCSI, 75
SDLC, 16, 37, 67, 187
SDM, 142
security, 198
segment bandwidth congestion, 19
segment switching, 224, 229
segmented backplane, 41
servers, 76, 266-268
ServerTrak, 121, 294
service levels, 59
SFS, 142
signal speed, 26
Simple Network Management Protocol
        (SNMP), 116-117
SimSoft, 291-293
simulation modeling, 147-149
Skylight, 118
slack time, 57
SLALOM, 141
SLDC, 7, 73
SLIP, 155
SmartDrive, 240
SMC, 221, 223
SMDS, 70
SMF, 118, 123
SNA, 7, 16, 37, 50, 63, 67, 73, 88, 90, 91, 118,
        132, 171, 187, 220, 245-246, 272
    gateway performance optimization, 257-259
SNMP, 22, 97, 110, 116-117, 120, 137, 220, 228
SofTrack, 123
SoftStar Job Scheduler, 122
software, component, 124
software management, 94
software metering, 123-124
SONET, 70, 115, 191, 192, 194, 212, 213
SPARCstation, 148
SPEC, 141, 142
SPEC series, 93
SpeedDisk, 240
SPM, 134
spoofing, 245
    definition, 81
SQL, 121-122, 239, 247, 249-253
SRB, 220
Stacker, 118, 268
state-change modeling, 147-148
statistical modeling, 150-153

STDA, 7, 206
store-and-forward switching, 226
StreetTalk, 206
Sun, 127
Sun NFS, 4
SunOS, 4
SVC, 214
SVR4, 66
switches and switching, 42-43, 76, 84-86
    ATM, 70
    matrix, 84
    network, 84
    packet, 42, 47, 224
    PBX, 38, 42
    port, 221, 229
    segment, 224, 229
    store-and-forward, 226
    virtual, 84, 223-229
    vs. routers, 160-161
SwitchIt, 226
Sybase, 247, 249
Sybase SQL, 122
synchronization, definition, 260
SynOptics, 41
SynOptics 5000, 229
SynOptics SNA workstation controller, 291
SYSCON, 145
SYSDEF, 122
SysDraw, 35, 107, 109, 297
system development multitasking (SDM),
        142
system file server (SFS), 142
system management facility (SMF), 123
SYSTEM.INI, 234
SYSUSE, 119

T

tank circuits, 77
TAPI, 97, 203
TCNS, 211
TCP/IP, 37, 88, 90, 91, 116, 118, 149, 213,
        219, 235, 236, 245
TCP/IP drivers, 236-237
TDM, 183-184, 204
telecommuting, 24
telephony, 97, 203
Telephony API (TAPI), 97, 203
throughput, 39, 50, 59, 68, 74, 239
THT, 47, 74
TIA/EIA, 131-132
time division multiplexing (TDM), 183-184, 204
TIMEOUT, 123
Token-Ring, 16, 36, 40, 47, 61, 74, 77, 81, 88,
        90, 132, 211, 221, 225
tools, 101-172
    performance, 102-103
TPC-A, 93, 141
TPC-B, 93, 141
TPC-C, 93, 142
TPC-D, 93, 142

TPS, 59, 121, 140, 171
TRACE, 121
traffic, 12-24
  cross-product, 171-172
  dynamics of, 13-23
  gridlock, 18-19, 22, 81
  patterns, 32
  peaks and bursts, 29-33
  prioritization, 16-17
  sources of, 71
  types demanding more bandwidth, 73
  volume, 33-34, 38-39
transaction modeling, LANModel and, 171
transaction processing, 50-51
  client/server, 248-266
  optimizing, 246-247
transaction processing load (TPS), 171
translation, definition, 81
transmission, full-duplex, 229-230
TrendTrak, 294
TRT, 47, 74
TRUSS, 121
TTY, 120
TUNESH, 120
tuning tools, 145
Tuxedo Enterprise Transaction Processing
    System, 121

U
UART, 95, 276
UDP, 218
UniCenter, 123
uninterruptible power supply (UPS), 231
UNIX, 4, 7, 96, 118, 120, 121, 122, 124, 125,
    126, 132, 134, 235, 240, 247, 268, 272
updating, 130
UPS, 231

V
VAR, 95
variable bit rate (VBR), 231
VAX, 118
VB Compress, 129
VBR, 231
VCFC, 231
VESA, 75
video, 98
Vines, 4, 7, 39, 90, 110, 115, 118, 139, 206,
    235, 276
Virtual Basic (VB), 253-255
virtual circuit flow control (VCFC), 231
virtual switch, 84, 223-229
Visual Basic, 128, 243, 247
Vital Signs, 118
VL, 75
VLAN, 227
VMS, 4, 115, 118, 235, 240, 272
VMSTAT, 119, 120, 122

voice, 97-98
voice mail, 202
volume, 33-34, 38-39
VSAT, 48, 65
VTAM, 118

W
WAN (see wide area network)
Western Union, 275
Whetstones, 141
wide area network (WAN), 46, 60, 148, 177,
    183, 191, 199, 211, 217, 246
  bandwidth, 46
  optimizing connections, 274-275
  optimizing, 234-236
Win, What, Where, 144, 303-304
WIN.INI, 234
Windows, 110, 118, 124, 126, 265
Windows for Workgroups, 240
Windows NT, 121, 240, 247
WinFaxPro, 179
WinSense, 115, 118
WinSock, 236
WinWord, 93
wire speed, 26
wiring errors in, 61
  infrastructure, 196
  structured, 176
wiring concentrator, 77, 78, 185
wiring route map, 104-105
word processing, 199, 234
workload, reducing, 205-206
workload characterization, 48-49, 74, 76, 144
workload contribution, 48

X
X Client, 272
X performance characterization (XPC), 143
X Server, 272
X Windows, 247, 272
X.25, 104, 118, 246
X.400, 7
X.500, 235
X/Open, 136
X/Open Portability Guide (XPG), 137
X11Perf, 143
Xbench, 143
Xmark, 143
XPC, 143
XPG, 137

Y
YACC, 130

Z
Zilog Z-80, 64

328   Index

CD-ROM WARRANTY

This software is protected by both United States copyright law and international copyright treaty provision. You must treat this software just like a book. By saying "just like a book," McGraw-Hill means, for example, that this software may be used by any number of people and may be freely moved from one computer location to another, so long as there is no possibility of its being used at one location or on one computer while it also is being used at another. Just as a book cannot be read by two different people in two different places at the same time, neither can the software be used by two different people in two different places at the same time (unless, of course, McGraw-Hill's copyright is being violated).

LIMITED WARRANTY

McGraw-Hill takes great care to provide you with top-quality software, thoroughly checked to prevent virus infections. McGraw-Hill warrants the physical diskette(s) contained herein to be free of defects in materials and workmanship for a period of sixty days from the purchase date. If McGraw-Hill receives written notification within the warranty period of defects in materials or workmanship, and such notification is determined by McGraw-Hill to be correct, McGraw-Hill will replace the defective diskette(s). Send requests to:

> McGraw-Hill, Inc.
> Customer Services
> P.O. Box 544
> Blacklick, OH 43004-0545

The entire and exclusive liability and remedy for breach of this Limited Warranty shall be limited to replacement of defective diskette(s) and shall not include or extend to any claim for or right to cover any other damages, including but not limited to, loss of profit, data, or use of the software, or special, incidental, or consequential damages or other similar claims, even if McGraw-Hill has been specifically advised of the possibility of such damages. In no event will McGraw-Hill's liability for any damages to you or any other person ever exceed the lower of suggested list price or actual price paid for the license to use the software, regardless of any form of the claim.

McGRAW-HILL, INC. SPECIFICALLY DISCLAIMS ALL OTHER WARRANTIES, EXPRESS OR IMPLIED, INCLUDING, BUT NOT LIMITED TO, ANY IMPLIED WARRANTY OF MERCHANTABILITY OR FITNESS FOR A PARTICULAR PURPOSE.

Specifically, McGraw-Hill makes no representation or warranty that the software is fit for any particular purpose and any implied warranty of merchantability is limited to the sixty-day duration of the Limited Warranty covering the physical CD-ROM only (and not the software) and is otherwise expressly and specifically disclaimed.

This limited warranty gives you specific legal rights; you may have others which may vary from state to state. Some states do not allow the exclusion of incidental or consequential damages, or the limitation on how long an implied warranty lasts, so some of the above may not apply to you.